REUSE METHODOLOGY MANUAL
FOR SYSTEM-ON-A-CHIP DESIGNS

REUSE METHODOLOGY MANUAL
FOR SYSTEM-ON-A-CHIP DESIGNS

by

Michael Keating
Synopsys, Inc.

and

Pierre Bricaud
Mentor Graphic Corporation

KLUWER ACADEMIC PUBLISHERS
Boston / Dordrecht / London

Distributors for North, Central and South America:
Kluwer Academic Publishers
101 Philip Drive
Assinippi Park
Norwell, Massachusetts 02061 USA

Distributors for all other countries:
Kluwer Academic Publishers
Distribution Centre
Post Office Box 322
3300 AH Dordrecht, THE NETHERLANDS

Library of Congress Cataloging-in-Publication Data

A C.I.P. Catalogue record for this book is available
from the Library of Congress.

Table of Contents

Foreword

The electronics industry has entered the era of multimillion-gate chips, and there's no turning back. By the year 2001, Sematech predicts that state-of-the-art ICs will exceed 12 million gates and operate at speeds surpassing 600 MHz. An engineer designing 100 gates/day would require a hypothetical 500 years to complete such a design, at a cost of $75 million in today's dollars. This will never happen, of course, because the time is too long and the cost is too high. But 12-million gate ICs will happen, and soon.

How will we get there? Whatever variables the solution involves, one thing is clear: the ability to leverage valuable intellectual property (IP) through design reuse will be the invariable cornerstone of any effective attack on the productivity issue. Reusable IP is essential to achieving the engineering quality and the timely completion of multimillion-gate ICs. Without reuse, the electronics industry will simply not be able to keep pace with the challenge of delivering the "better, faster, cheaper" devices consumers expect.

Synopsys and Mentor Graphics have joined forces to help make IP reuse a reality. One of the goals of our Design Reuse Partnership is to develop, demonstrate, and document a reuse-based design methodology that works. The *Reuse Methodology Manual* (RMM) is the result of this effort. It combines the experience and resources of Synopsys and Mentor Graphics. Synopsys' expertise in design reuse tools and Mentor Graphics' expertise in IP creation and sourcing resulted in the creation of this manual that documents the industry's first systematic reuse methodology. The RMM describes the design methodology that our teams have found works best for designing reusable blocks and for integrating reusable blocks into large chip designs.

It is our hope that this manual for advanced IC designers becomes the basis for an industry-wide solution that accelerates the adoption of reuse and facilitates the rapid development of tomorrow's large, complex ICs.

Aart J. de Geus   *Walden C. Rhines*
Chairman & CEO   *President & CEO*
Synopsys, Inc.   *Mentor Graphics Corporation*

Acknowledgements

We would like to thank the following people who make substantial contributions to the ideas and content of the *Reuse Methodology Manual*:

- Warren Savage, Ken Scott, and their engineering teams, including Shiv Chonnad, Guy Huchison, Chris Kopetsky, Keith Rieken, Mark Noll, and Ralph Morgan
- Glenn Dukes and his engineering team
- John Coffin, Ashwini Mulgaonkar, Suzanne Hayek, Pierre Thomas, Alain Pirson, Fathy Yassa, John Swanson, Gil Herbeck, Saleem Haider, Martin Lampard, Larry Groves, Norm Kelley, Kevin Kranen, Angelina So, and Neel Desai

We would also like to thank the following individuals for their helpful suggestions on how to make the RMM a stronger document:

- Nick Ruddick, Sue Dyer, Jake Buurma, Bill Bell, Scott Eisenhart, Andy Betts, Bruce Mathewson

Finally, we would like to thank Rhea Tolman and Bill Rogers for helping to prepare the manuscript.

CHAPTER 1   *Introduction*

Silicon technology now allows us to build chips consisting of tens of millions of transistors. This technology promises new levels of system integration onto a single chip, but also presents significant challenges to the chip designer. As a result, many ASIC developers and silicon vendors are re-examining their design methodologies, searching for ways to make effective use of the huge numbers of gates now available.

These designers see current design tools and methodologies as inadequate for developing million gate ASICs from scratch. There is considerable pressure to keep design team size and design schedules constant even as design complexities grow. Tools are not providing the productivity gains required to keep pace with the increasing gate counts available from deep submicron technology. Design reuse—the use of pre-designed and pre-verified cores—is the most promising opportunity to bridge the gap between available gate-count and designer productivity.

This manual outlines an effective methodology for creating reusable designs for use in a System-on-a-Chip (SoC) design methodology. Silicon and tool technologies move so quickly that no single methodology can provide a permanent solution to this highly dynamic problem. Instead, this manual is an attempt to capture and incrementally improve on current best practices in the industry, and to give a coherent, integrated view of the design process. We expect to update this document on a regular basis as a result of changing technology and improved insight into the problems of design reuse and its role in producing high-quality SoC designs.

1.1   Goals of This Document

Development methodology necessarily differs between system designers and ASSP designers, as well as between DSP developers and chipset developers. However, there is a common set of problems facing everyone who is designing SoC-scale ASICs:

- Time-to-market pressures demand rapid development.
- Quality of results, in performance, area, and power, are key to market success.
- Increasing chip complexity makes verification more difficult.
- The development team has different levels and areas of expertise, and is often scattered throughout the world.
- Design team members may have worked on similar designs in the past, but cannot reuse these designs because the design flow, tools, and guidelines have changed.
- SoC designs include embedded processor cores, and thus a significant software component, which leads to additional methodology, process, and organizational challenges.

In response to these problems, many design teams are turning to a block-based design approach that emphasizes design reuse. Reusing macros (sometimes called "cores") that have already been designed and verified helps to address all of the above problems. However, ASIC design for reuse is a new paradigm in hardware design. Ironically, many researchers in software design reuse point to hardware design as the prime model for design reuse, in terms of reusing the same chips in different combinations to create many different board designs. However, most ASIC design teams do not code their RTL or design their testbenches with reuse in mind and, as a result, most designers find it faster to develop modules from scratch than to reverse engineer someone else's design.

Some innovative design teams are trying to change this pattern and are developing effective design reuse strategies. This document focuses on describing these techniques. In particular, it describes:

- How reusable macros fit into a System-on-a-Chip development methodology
- How to design reusable soft macros
- How to design reusable hard macros
- How to integrate soft and hard macros into a System-on-a-Chip design
- How to verify timing and functionality in large System-on-a-Chip designs

In doing so, this document addresses the concerns of two distinct audiences: the creators of reusable designs (macro authors) and chip designers who use these reusable blocks (macro integrators). For macro authors, the main sections of interest will be those on how to design reusable hard and soft macros, and the other sections will be primarily for reference. For integrators, the sections on designing hard and soft macros are intended primarily as a description of what to look for in reusable designs.

System-on-a-Chip designs are made possible by deep submicron technology. This technology presents a whole set of design challenges. Interconnect delays, clock and power distribution, and place and route of millions of gates are real challenges to physical design in the deep submicron technologies. These physical design problems can have a significant impact on the functional design of systems on a chip and on the design process itself. Interconnect issues, floorplanning, and timing design must be addressed early in the design process, at the same time as the development of the functional requirements. This document addresses issues and problems related to providing logically robust designs that can be fabricated on deep submicron technologies and that, when fabricated, will meet the requirements for clock speed, power, and area.

System-on-a-Chip designs have a significant software component in addition to the hardware itself. However, this manual focuses primarily on the creation and reuse of reusable hardware macros. This focus on hardware reuse should not be interpreted as an attempt to minimize the importance in the software aspects of system design. Software plays an essential role in the design, integration, and test of SoC systems, as well as in the final product itself.

1.1.1   Assumptions

This document assumes that the reader is familiar with standard high-level design methodology, including:

- HDL design and synthesis
- Design for test, including full scan techniques
- Floorplanning and place and route

1.1.2   Definitions

In this document, we will use the following terms interchangeably:

- Macro
- Core
- Block

All of these terms refer to a design unit that can reasonably be viewed as a stand-alone sub-component of a complete System-on-Chip design. Examples include a PCI interface macro, a microprocessor core, or an on-chip memory.

Other terms used throughout this document include:

- **Subblock** – A subblock is a sub-component of a macro, core, or block. It is too small or specific to be a stand-alone design component.
- **Hard macro** – A hard macro (or core or block) is one that is delivered to the integrator as a GDSII file. It is fully designed, placed, and routed by the supplier.
- **Soft macro** – A soft macro (or core or block) is one that is delivered to the integrator as synthesizable RTL code.

1.1.3   Virtual Socket Interface Alliance

The Virtual Socket Interface Alliance (VSIA) is an industry group working to facilitate the adoption of design reuse by setting standards for tool interfaces and design practices. VSIA has done an excellent job in raising industry awareness of the importance of reuse and of identifying key technical issues that must be addressed to support widespread and effective design reuse.

The working groups of the VSIA have developed a number of proposals for standards that are currently in review. To the extent that detailed proposals have been made, this document attempts to be compliant with them.

Some exceptions to this position are:

- Virtual component: VSIA has adopted the name "virtual component" to specify reusable macros. We have used the shorter term "macro" in most cases.
- Firm macro: VSIA has defined an intermediate form between hard and soft macros, with a fairly wide range of scope. Firm macros can be delivered in RTL or netlist form, with or without detailed placement, but with some form of physical design information to supplement the RTL itself. We do not address firm macros specifically in this document; we feel that it is more useful to focus on hard and soft macros. As technology evolves for more tightly coupling synthesis and physical design, we anticipate that the category of firm macros will be merged with that of soft macros.

1.2   Design for Reuse: The Challenge

An effective block-based design methodology requires an extensive library of reusable blocks, or macros. The developers of these macros must, in turn, employ a design methodology that consistently produces reusable macros. This design reuse methodology is based on the following principles:

- Creation of every stage of design, from specification to silicon, with the understanding that it will be modified and reused in other projects by other design teams

- The use of tools and processes that capture the design information in a consistent, easy-to-communicate form
- The use of tools and processes that make it easy to integrate modules into a design when the original designer is not available

1.2.1 Design for Use

Design for reuse presents significant new challenges to the design team. But before considering innovations, remember that to be *reusable*, a design must first be *usable*: a robust and correct design. Many of the techniques for design reuse are just good design techniques:

- Good documentation
- Good code
- Thorough commenting
- Well-designed verification environments and suites
- Robust scripts

Both hardware and software engineers learn these techniques in school, but in the pressures of a real design project, they often succumb to the temptation to take short-cuts. A shortcut may appear to shorten the design cycle for code that is used only once, but it often prevents the code from being effectively reused by other teams on other projects. Initially, complying with these design reuse practices might seem like an extra burden, but once the design team is fully trained, these techniques speed the design, verification, and debug processes of a project by reducing iterations through-out the code and verification loop.

1.2.2 Design for Reuse

In addition to the requirements above for a robust design, there are some additional requirements for a hardware macro to be fully reusable. The macro must be:

- **Designed to solve a general problem** – This often means the macro must be easily configurable to fit different applications.
- **Designed for use in multiple technologies** – For soft macros, this means that the synthesis scripts must produce satisfactory quality of results with a variety of libraries. For hard macros, this means having an effective porting strategy for mapping the macro onto new technologies.
- **Designed for simulation with a variety of simulators** – A macro or a verification testbench that works with only a single simulator is not portable. Some new simulators support both Verilog and VHDL. However, good design reuse

practices dictate that both a Verilog and VHDL version of each model and verification testbench should be available, and they should work with all the major commercial simulators.

- **Verified independently of the chip in which it will be used** – Often, macros are designed and only partially tested before being integrated into a chip for verification, thus saving the effort of developing a full testbench for the design. Reusable designs must have full, stand-alone testbenches and verification suites that afford very high levels of test coverage.

- **Verified to a high level of confidence** – This usually means very rigorous verification as well as building a physical prototype that is tested in an actual system running real software.

- **Fully documented in terms of appropriate applications and restrictions** – In particular, valid configurations and parameter values must be documented. Any restrictions on configurations or parameter values must be clearly stated. Interfacing requirements and restrictions on how the macro can be used must be documented.

These requirements increase the time and effort needed for the development of a macro, but they provide the significant benefit of making that macro reusable.

1.2.3   Fundamental Problems

Teams attempting to reuse code today are frequently faced with code that wasn't designed for reuse. The guidelines and techniques described in this document are the result of our experience with problems, such as:

- The design representation is not appropriate. For example, the RTL is available in Verilog but the new chip design is in VHDL, or a gate-level netlist using a .5μ library is available, but an incompatible .35μ library is now being used.

- The design comes with incomplete design information, often with no functional specification and with unreadable, uncommented code.

- Supporting scripts are not available or are so obtuse as to be unusable.

- The full design was never properly archived, so pieces of the design are scattered over various disks on various machines, some of which no longer exist.

- The tools used to develop the design are no longer supported; vendors have gone out of business.

- The tools used to develop the design had poor inter-operability; scripts to patch the tools together have disappeared.

- A hard macro is available, but the simulation model is so slow that system level simulation is not practical.

CHAPTER 2

The System-on-a-Chip Design Process

This chapter gives an overview of the System-on-a-Chip (SoC) design methodology. The topics include:

- Canonical SoC design
- SoC design flow
- The role of specifications throughout the life of a project

2.1 A Canonical SoC Design

Consider the chip design in Figure 2-1. We claim that, in some sense, this design represents the canonical or generic form of System-on-a-Chip design. It consists of:

- A microprocessor and its memory subsystem
- A datapath that includes interfaces to the external system
- Blocks that perform transformations on data received from the external system
- Another I/O interface to the external system

This design is somewhat artificial, but it contains most of the structures and challenges found in real SoC designs. The processor could be anything from an 8-bit 8051 to a 64-bit RISC. The memory subsystem could be single or multi-leveled, and could include SRAM and/or DRAM. The communication interfaces could include PCI, Ethernet, USB, A-to-D, D-to-A, electro-mechanical, or electo-optical converters. The data transformation block could be a graphics processor or a network router. The design process required to specify such a system, to develop and verify the blocks, and to assemble them into a fabricated chip contains all the basic elements and challenges of a SoC design.

Real SoC designs are, of course, much more complex than this canonical example. A real design would typically include several sets of IP interfaces and data transformations. Many SoC designs today include multiple processors, and combinations of processors and DSPs. The memory structures of SoC designs are often very complex as well, with various levels of caching and shared memory, and specific data structures to support data transformation blocks, such as MPEG2. Thus, the canonical design is just a miniature version of a SoC design that allows us to discuss the challenges of developing these chips utilizing reusable macros.

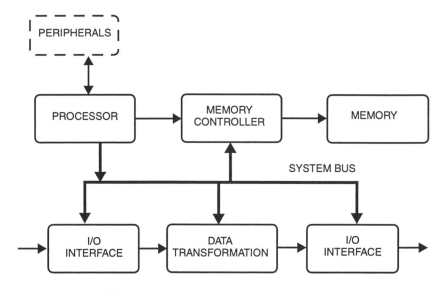

Figure 2-1   Canonical hardware view of SoC

2.2   System Design Flow

To meet the challenges of SoC, chip designers are changing their design flows in two major ways:

• From a waterfall model to a spiral model
• From a top-down methodology to a combination of top-down and bottom-up

2.2.1   Waterfall vs. Spiral

The traditional model for ASIC development, shown in Figure 2-2, is often called a *waterfall model*. In a waterfall model, the project transitions from phase to phase in a step function, never returning to the activities of the previous phase. In this model, the design is often tossed "over the wall" from one team to the next without much interaction between the teams.

This process starts with the development of a specification for the ASIC. For complex ASICs with high algorithmic content, such as graphics chips, the algorithm may be developed by a graphics expert; this algorithm is then given to a design team to develop the RTL for the ASIC.

After functional verification, either the design team or a separate team of synthesis experts synthesizes the ASIC into a gate-level netlist. Then timing verification is performed to verify that the ASIC meets timing. Once the design meets its timing goals, the netlist is given to the physical design team, which places and routes the design. Finally, a prototype chip is built and tested.

This flow has worked well in designs of up to 100k gates and down to .5 µ. It has consistently produced chips that worked right the first time, although often the systems that were populated with them did not.

For large, deep submicron designs, this methodology simply does not work. Large systems have sufficient software content that hardware and software must be developed concurrently to ensure correct system functionality. Physical design issues must be considered early in the design process to ensure that the design can meet its performance goals.

As complexity increases, geometry shrinks, and time-to-market pressures continue to escalate, chip designers are turning to a modified flow to produce today's larger SoC designs. Many teams are moving from the old waterfall model to the newer *spiral development model*. In the spiral model, the design team works on multiple aspects of the design simultaneously, incrementally improving in each area as the design converges on completion.

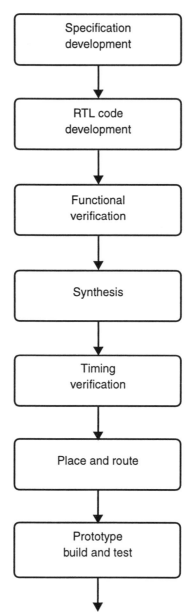

Figure 2-2   Traditional ASIC design flow

Figure 2-3 shows the SoC design flow. This flow is characterized by:

- Parallel, concurrent development of hardware and software
- Parallel verification and synthesis of modules
- Floorplanning and place-and-route included in the synthesis process
- Modules developed only if a predesigned hard or soft macro is not available
- Planned iteration throughout

In the most aggressive projects, engineers simultaneously develop top-level system specifications, algorithms for critical subblocks, system-level verification suites, and timing budgets for the final chip integrations. That means that they are addressing all aspects of hardware and software design concurrently: functionality, timing, physical design, and verification.

2.2.2 Top-Down vs. Bottom-Up

The classic top-down design process can be viewed as a recursive routine that begins with specification and decomposition, and ends with integration and verification:

1. Write complete specifications for the system or subsystem being designed.
2. Refine its architecture and algorithms, including software design and hardware/software cosimulation if necessary.
3. Decompose the architecture into well-defined macros.
4. Design or select macros; this is where the recursion occurs.
5. Integrate macros into the top level; verify functionality and timing.
6. Deliver the subsystem/system to the next higher level of integration; at the top level, this is tapeout.
7. Verify all aspects of the design (functionality, timing, etc.).

With increasing time-to-market pressures, design teams have been looking at ways to accelerate this process. Increasingly powerful tools, such as synthesis and emulation tools, have made significant contributions. Developing libraries of reusable macros also aids in accelerating the design process.

A top-down methodology assumes that the lowest level blocks specified can, in fact, be designed and built. If it turns out that a block is not feasible to design, the whole specification process has to be repeated. For this reason, real world design teams usually use a mixture of top-down and bottom-up methodologies, building critical low-level blocks while they refine the system and block specifications. Libraries of reusable hard and soft macros clearly facilitate this process by providing a source of pre-verified blocks, proving that at least some parts of the design can be designed and fabricated in the target technology and perform to specification.

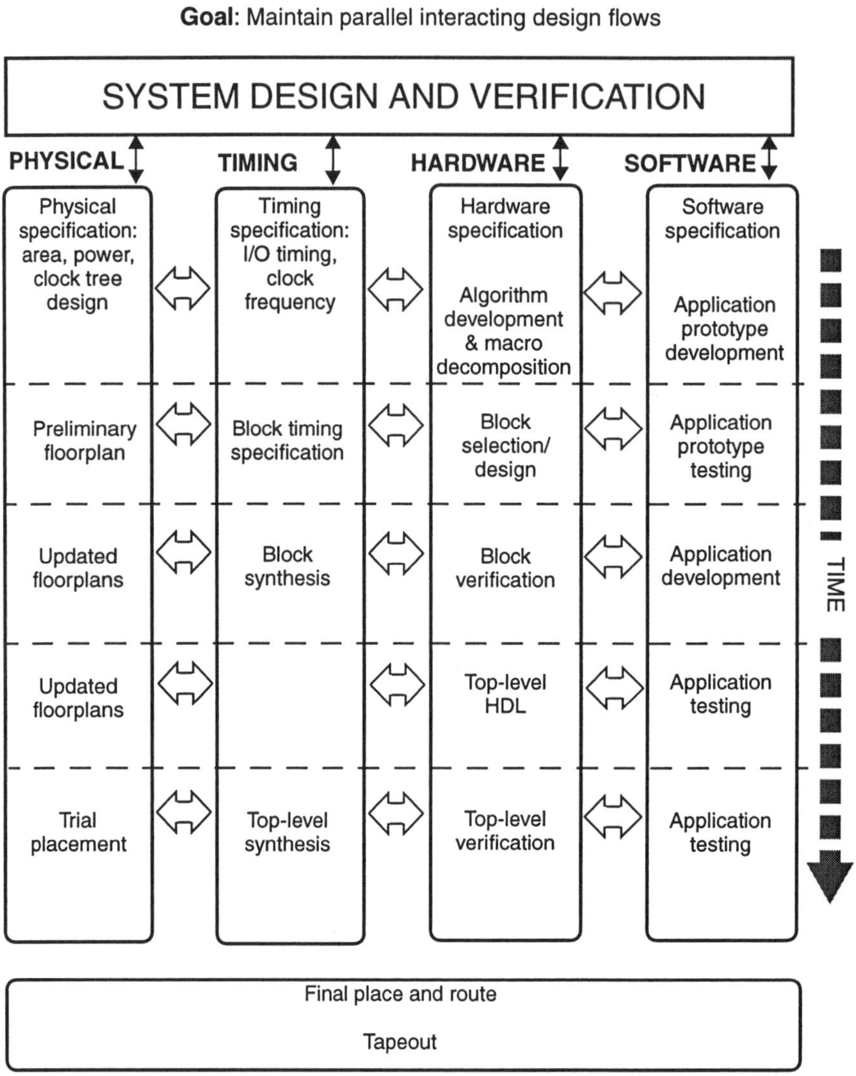

Goal: Maintain parallel interacting design flows

SYSTEM DESIGN AND VERIFICATION			
PHYSICAL	**TIMING**	**HARDWARE**	**SOFTWARE**
Physical specification: area, power, clock tree design	Timing specification: I/O timing, clock frequency	Hardware specification Algorithm development & macro decomposition	Software specification Application prototype development
Preliminary floorplan	Block timing specification	Block selection/ design	Application prototype testing
Updated floorplans	Block synthesis	Block verification	Application development
Updated floorplans		Top-level HDL	Application testing
Trial placement	Top-level synthesis	Top-level verification	Application testing

TIME

Final place and route

Tapeout

Figure 2-3   System-on-a-Chip design flow

2.2.3 Construct by Correction

The Sun Microsystems engineers that developed the UltraSPARC processor have described their design process as "construct by correction." In this project, a single team took the design from architectural definition through place and route. In this case, the engineers had to learn how to use the place and route tools, whereas, in the past, they had always relied on a separate team for physical design. By going through the entire flow, the team was able to see for themselves the impact that their architectural decisions had on the area, power, and performance of the final design.

The UltraSPARC team made the first pass through the design cycle—from architecture to layout—as fast as possible, allowing for multiple iterations through the entire process. By designing an organization and a development plan that allowed a single group of engineers to take the design through multiple complete iterations, the team was able to see their mistakes, correct them, and refine the design several times before the chip was finally released to fabrication. The team called this process of iteration and refinement "construct by correction".

This process is the opposite of "correct by construction" where the intent is to get the design completely right during the first pass. The UltraSPARC engineers believed that it was not possible at the architectural phase of the design to foresee all the implication their decisions would have on the final physical design.

The UltraSPARC development projects was one of the most successful in Sun Microsystems' history. The team attributes much of its success to the "construct by correction" development methodology.

2.3 The Specification Problem

The first part of the design process consists of recursively developing, verifying, and refining a set of specifications until they are detailed enough to allow RTL coding to begin. The rapid development of clear, complete, and consistent specifications is a difficult problem. In a successful design methodology, it is the most crucial, challenging, and lengthy phase of the project. If you know what you want to build, implementation mistakes are quickly spotted and fixed. If you don't know, you may not spot major errors until late in the design cycle or until fabrication.

Similarly, the cost of documenting a specification during the early phases of a design is much less than the cost of documenting it after the design is completed. The extra discipline of formalizing interface definitions, for instance, can occasionally reveal inconsistencies or errors in the interfaces. On the other hand, documenting the design after it is completed adds no real value for the designer and either delays the project or is skipped altogether.

2.3.1   Specification Requirements

In a SoC design, specifications are required for both the hardware and software portions of the design. The specifications must address:

Hardware

- Functionality
- Timing
- Performance
- Interface to SW
- Physical design issues such as area and power

Software

- Functionality
- Timing
- Performance
- Interface to HW
- SW structure, kernel

Traditionally, specifications have been written in a natural language, such as English, and have been plagued by ambiguities, incompleteness, and errors. Many companies, realizing the problems this causes have started using executable specifications for some or all of the system.

2.3.2   Types of Specifications

There are two major techniques currently being used to help make hardware and software specifications more robust and useful: *formal specification* and *executable specification*.

- **Formal specification** – In formal specification, the desired characteristics of the design are defined independently of any implementation. This type of specification is considered promising in the long term. Once a formal specification is generated for a design, formal methods such as property checking can be used to prove that a specific implementation meets the requirements of the specification. A number of formal specification languages have been developed, including one for VHDL called VSPEC. These languages typically provide a mechanism for describing not only functional behavior, but timing, power, and area requirements as well. To date, formal specification has not been used widely for commercial designs, but continues to be an important research topic.

- **Executable specifications** – Executable specifications are currently more useful for describing functional behavior in most design situations. An executable specification is typically an abstract model for the hardware and/or software being specified. It is typically written in C, C++, or SDL for high level specifications; at the lower levels, hardware is usually described in Verilog or VHDL. Developing these software models early in the design process allows the design team to verify the basic functionality and interfaces of the hardware and software long before the detailed design begins.

  Most executable specifications address only the functional behavior of a system, so it may still be necessary to describe critical physical specifications—timing, clock frequency, area, and power requirements—in a written document. Efforts are under way to develop more robust forms of capturing timing and physical design requirements.

2.4   The System Design Process

The system design process shown in Figure 2-4 employs both executable and written specifications. This process involves the following steps:

1. **System specification**

   The process begins by identifying the *system requirements*: the required functions, performance, cost, and development time for the system. These are formulated into a *preliminary specification*, often written jointly by engineering and marketing. Then, a *high-level algorithmic model* for the overall system is developed, usually in C/C++. Tools such as COSSAP, SPW, and Matlab may be more useful for some algorithmic-intensive designs, and tools such as Bones, NuThena, SDT more useful for control dominated designs.

   This high-level model provides an executable specification for the key functions of the system.

2. **Model refinement and test**

   A verification environment for the high-level model is developed to *refine and test* the algorithm. This environment provides a mechanism for refining the high-level design, verifying the functionality and performance of the algorithm. If properly designed, it can also be used later to verify models for the hardware and software, such as an RTL model verified using hardware/software cosimulation. For systems with very high algorithmic content, considerable model development, testing, and refinement occurs before the hardware/software partitioning. The focus here is on the algorithm, not the implementation.

3. Hardware/software partitioning (decomposition)

As the high-level model is refined, the system architects determine the *hardware/software partition*; that is, the division of system functionality between hardware and software. This is largely a manual process requiring judgment and experience on the part of the system architects and a good understanding of the cost/performance trade-offs for various architectures. A rich library of preverified, characterized macros and a rich library of reusable software modules are essential for identifying the size and performance of various hardware and software functions. Tools, such as NuThena's Forsight can assist in the validation and performance estimates of a partition.

The final step in hardware/software partitioning is to define the interfaces between hardware and software, and specify the communication protocols between them.

4. Block specification

The output of this partitioning is a *hardware specification* and a *software specification*. The hardware specification includes a description of the basic functions, the timing, area, and power requirements, and the physical and software interfaces, with detailed descriptions of the I/O pins and the register map.

5. System behavioral model and cosimulation

Once the hardware/software partition is determined, a behavioral model of the hardware is developed in parallel with a prototype version of the software. Often these can be derived from the system model and from behavioral models of hardware functions that already exist in a library of macros. Hardware/software cosimulation then allows the hardware model and prototype software to be refined to the point where a robust executable and written functional specification for each is developed. This hardware/software cosimulation continues throughout the design process, verifying interoperability between the hardware and software at each stage of design.

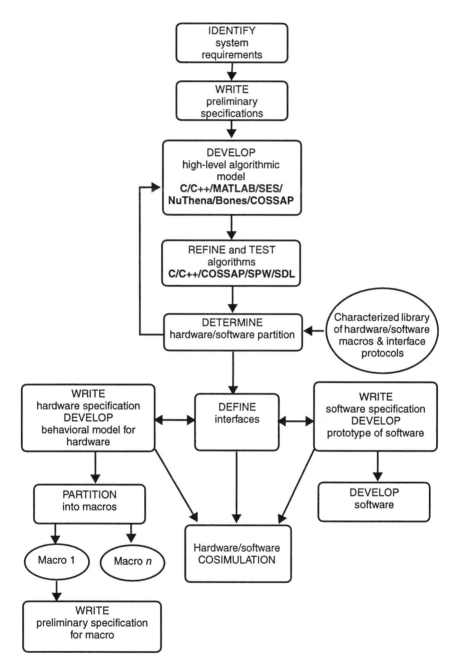

Figure 2-4   Top-level system design and recommended applications for each step

<table>
<tr><td>CHAPTER 3</td><td></td></tr>
</table>

CHAPTER 3   *System-Level Design Issues: Rules and Tools*

This chapter discusses system-level issues such as layout, clocking, floorplanning, on-chip busing, and strategies for synthesis, verification, and testing. These elements must be agreed upon *before* the components of the chip are selected or designed.

Topics in this chapter include:

- Interoperability issues
- Timing and synthesis issues
- Functional design issues
- Physical design issues
- Verification strategy
- Manufacturing test strategies

3.1   Interoperability Issues

Two of the biggest issues affecting the success of a SoC design are the interoperability of the macros being integrated and the interoperability of the tools used to implement the design.

It is essential that the design team agree to a set of design rules for the entire chip before beginning any design or macro selection. By agreeing in advance on such critical issues as clocking and reset strategy, macro interface architecture, and design for test, the team can select and design macros that will work well together.

This chapter describes the basic set of design guidelines that should be explicitly agreed to by the design team before starting implementation or macro selection. Most of these requirements are aimed at facilitating rapid chip-level integration and verification of the various macros.

3.2   Timing and Synthesis Issues

Timing and synthesis issues include synchronous or asynchronous design, clock and reset schemes, and selection of synthesis strategy.

3.2.1   Synchronous vs. Asynchronous Design Style

Rule – The system should be synchronous and register based. Latches should be used only to implement small memories or FIFOs. The FIFOs and memories should be designed so that they are synchronous to the external world and are edge triggered. Exceptions to this rule should be made with great care and must be fully documented.

In the past, latch-based designs have been popular, especially for some processor designs. Multi-phase, non-overlapping clocks were used to clock the various pipeline stages. Latches were viewed as offering greater density and higher performance than register (flop) based designs. These benefits were sufficient to justify the added complexity of design.

Today, the tradeoffs are quite different. Deep submicron technology has made a huge number of gates available to the chip designer and, in most processor-based designs, the size of on-chip memory is dwarfing the size of the processor pipeline. Also, with deep submicron design, delays are dominated by interconnect delay, so the difference in effective delay between latches and flip-flops is minimal.

On the other hand, the cost of the increased complexity of latch-based design has risen significantly with the increase in design size and the need for design reuse.

Latch timing is inherently ambiguous, as illustrated in Figure 3-1. The designer may intend data to be set up at the D input of the latch before the leading edge of the clock, in which case data is propagated to the output on the leading edge of clock. Or, the designer may intend data to be set up just before the trailing edge of the clock, in which case data is propagated to the output (effectively) on the trailing edge of the clock.

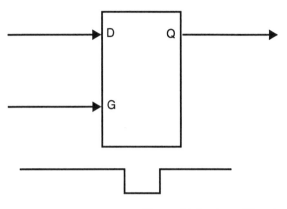

From which edge of the clock is
data propagated to the output?

Figure 3-1   Latch timing

Designers may take advantage of this ambiguity to improve timing. "Time borrow-ing" is the practice of absorbing some delay by:

- Guaranteeing that the data is set up before the leading clock edge at one stage
- Allowing data to arrive as late as one setup time before the trailing clock edge at the next stage

The problem caused by the ambiguity of latch timing, and exacerbated by time bor-rowing, is that it is impossible by inspection of the circuit to determine whether the designer intended to borrow time or the circuit is just slow. Thus, timing analysis of each latch of the design is difficult. Over a large design, timing analysis becomes impossible. Only the original designer knows the full intent of the design. Thus, latch-based design is inherently not reusable.

For this reason, true latch-based designs are not appropriate for SoC designs. Some LSSD design styles are effectively register-based and are acceptable if used correctly.

3.2.2   Clocking

Rule – The number of clock domains and clock frequencies must be documented. It is especially important to document:

- Required clock frequencies and associated phase locked loops
- External timing requirements (setup/hold and output timing) needed to interface to the rest of the system

Guideline – Use the smallest possible number of clock domains. If two asynchronous clock domains interact, they should meet in a single module, which should be as small as possible. Ideally, this module should consist solely of the flops required to transfer the data from one clock domain to the other. The interface structure between the two clock domains should be designed to avoid metastability.

Guideline – If a phase locked loop (PLL) is used for on-chip clock generation, then some means of disabling or bypassing the PLL should be provided. This bypass makes chip testing and debug much easier, and facilitates using hardware modelers for system simulation.

3.2.3   Reset

Rule – The basic reset strategy for the chip must be documented. It is particularly important to address the following issues:

- Is it synchronous or asynchronous?
- Is there an internal or external power-on reset?
- Is there more than one reset (hard vs. soft reset)?
- Is each macro individually resettable for debug purposes?

There are advantages and disadvantages to both synchronous and asynchronous reset.

Synchronous reset:

- Is easy to synthesize — reset is just another synchronous input to the design.
- Requires a free-running clock, especially at power-up, for reset to occur.

Asynchronous reset:

- Is harder to synthesize — reset is a special signal, like clock. Usually, a tree of buffers is inserted at place and route.
- Does not require a free-running clock.
- Must be synchronously de-asserted in order to ensure that all flops exit the reset condition on the same clock. Otherwise, state machines can reset into invalid states.

Guideline – Synchronous reset is preferred, because the timing model for it is much simpler than for asynchronous reset. Synchronous reset avoids possible race conditions on reset that are virtually impossible to simulate, and that make static timing analysis more difficult. With asynchronous reset, the designer needs to worry about pulse width throughout the circuit, as well as making sure that reset is de-asserted synchronously to the clock.

3.2.4   Synthesis Strategy and Timing Budgets

Rule – Overall design goals for timing, area, and power should be documented before macros are designed or selected. In particular, the overall chip synthesis methodology needs to be planned very early in the chip design process.

We recommend a bottom-up synthesis approach. Each macro should have its own synthesis script, that ensures that the internal timing of the macro can be met in the target technology. This implies that the macro should be floorplanned as a single unit to ensure that the original wire load model still holds and is not subsumed into a larger floorplanning block.

Chip-level synthesis then consists solely of connecting the macros and resizing output drive buffers to meet actual wire load and fanout. To facilitate this, the macro should appear at the top level as two blocks: the internals of the macro (which are `dont_touched`) and the output buffers (which undergo incremental compile).

3.3   Functional Design Issues

Functional design issues include system interconnect and debug strategy.

3.3.1   System Interconnect and On-Chip Buses

Rule – The design of the on-chip busing scheme that will interconnect the various blocks in a SoC design must be an integral part of the macro selection and design process. If it is done after the fact, conflicting bus designs are likely to require additional interface hardware design and could jeopardize system performance.

Guideline – There are different bus strategies for different kinds of blocks used in a SoC design. Microprocessors and microcontrollers tend to have fixed interfaces, so it is necessary to design or select peripherals that can interface to the selected micro-controller.

Because of the need to interface to a variety of buses, it is best to design or select macros that have flexible or parameterizable interfaces. FIFO-based interfaces are particularly flexible; they have simple interfaces, simple timing requirements, and can compensate for different data rates between the macro and the bus.

The PI-Bus defined by the Open Microprocessor Systems Initiative (OMI), and the FISPbus from Mentor Graphics, are examples of on-chip buses. We believe most on-chip buses will share many of the characteristics of these standards, including:

- Separate address and data buses
- Support for multiple masters
- Request/grant protocol
- Fully synchronous, multiple-cycle transactions

3.3.2   Debug Structures

Rule – The design team must develop a strategy for the bring-up and debug of the SoC design at the beginning of the design process. The most effective debug strategies usually require specific features to be designed into the chip. Adding debug features early in the design cycle greatly reduces the incremental cost of these features, in terms of design effort and schedule. Adding debug features after the basic functionality is designed can be difficult or impossible, and can be very time consuming. However, without effective debug structures, even the simplest of bugs can be very difficult to troubleshoot on a large SoC design.

Guideline – Controllability and observability are the keys to an easy debug process.

- *Controllability* is best achieved by design features in the macros themselves. The system should be designed so that each macro can be effectively turned off, turned on, or put into a debug mode where only its most basic functions are operational. This can be done either from an on-chip microprocessor or microcontroller, or from the chip's test controller.
- *Observability* is best achieved by adding bus monitors to the system. These monitors check data transactions, detect illegal transactions, and provide a logic analyzer type of interface to the outside world for debugging.

3.4   Physical Design Issues

Physical design issues include integrating hard and soft macros, floorplanning, and clock distribution.

3.4.1   Hard Macros

Rule – A strategy for floorplanning, placing, and routing a combination of hard and soft macros must be developed *before* hard macros are selected or designed for the chip. Most SoC designs combine hard and soft macros, and hard macros are problematic because they can cause blockage in the placement and routing of the entire chip. Too many hard macros, or macros with the wrong aspect ratio, can make the chip unroutable or unacceptably big, or can create unacceptable delays on critical nets.

3.4.2   Floorplanning

Rule – As a corollary to the previous rule, floorplanning must begin early in the design process. The size of the chip is critical in determining whether the chip will meet its timing, performance, and cost goals. Some initial floorplan should be developed as part of the initial functional specification for the SoC design.

This initial floorplan can be critical in determining both the functional interfaces between macros and the clock distribution requirements for the chip. If macros that communicate with each other must be placed far apart, signal delays between the macros may exceed a clock cycle, forcing a lower-speed interface between the macros.

3.4.3   Clock Distribution

Rule – The design team must decide on the basic clock distribution architecture for the chip early in the design process. The size of the chip, the target clock frequency, and the target library are all critical in determining the clock distribution architecture.

To date, most design teams have used a balanced clock tree to distribute a single clock throughout the chip, with the goal of distributing the clock with a low enough skew to prevent hold-time violations for flip-flops that directly drive other flip-flops.

For large, high-speed chips, this approach can require extremely large, high-power clock buffers. These buffers can consume as much as half of the power in the chip and a significant percentage of the real estate.

Guideline – For chips attempting to achieve lower power consumption, design teams are turning to a clock distribution technique similar to that used on boards today. A lower-speed bus is used to connect the modules and all transactions between modules use this bus. The bus is fully synchronous and a clock is distributed as one of the bused signals. The clock distribution for this bus still requires relatively low skew, but the distribution points for the clock are much fewer. Each macro can then synchronize its own local clock to the bus clock, either by buffering the bus clock or by using a phase locked loop. This local clock can be a multiple of the bus clock, allowing higher frequency clocking locally.

3.5   Verification Strategy

Rule – The system level verification strategy must be developed and documented before macro selection or design begins. Selecting or designing a macro that does not provide the modeling capability required for system-level verification can prevent otherwise successful SoC designs from completing in a timely manner. See Chapter 11 for a detailed discussion of verification strategies.

Guideline – The verification strategy determines which verification tools can be used. These tools could include event-driven simulation, cycle-based simulation, and/or emulation. Each of these tools could have very specific requirements in terms of coding style. If a required macro or testbench is not coded in the style required by the tool, the design team may have to spend a significant amount of effort to translate the code.

The verification strategy also determines the kinds of testbenches required for system-level verification. These testbenches must accurately reflect the environment in which the final chip will work, or else we are back in the familiar position of "the chip works, but the system doesn't." Testbench design at this level is non-trivial and must be started early in the design process.

3.6   Manufacturing Test Strategies

Manufacturing test strategies must be established at specification time. The optimal strategy for an individual block depends on the type of block.

3.6.1   System Level Test Issues

Rule – The system-level chip manufacturing test strategy must be documented.

Guideline – On-chip test structures are recommended for all blocks. It is not feasible to develop parallel test vectors for chips consisting of over a million gates. Different kinds of blocks will have different test strategies; at the top level, a master test controller is required to control and sequence these independent test structures.

3.6.2   Memory Test

Guideline – Some form of BIST is recommended for RAMs, because this provides a rapid, easy-to-control test methodology. However, some BIST solutions are not sufficient to test data retention. Some form of reasonably direct memory access is recommended to detect and troubleshoot data retention problems.

3.6.3   Microprocessor Test

Guideline – Microprocessors usually have some form of custom test structure, combining full or partial scan and parallel vectors. Often, this means that the chip-level test controller must provide the microprocessor with both a scan chain controller and some form of boundary scan.

3.6.4   Other Macros

Guideline – For most other blocks, the best choice is a full-scan technique. Full scan provides very high coverage for very little design effort. The chip-level test controller needs to manage the issue of how many scan chains are operated simultaneously, and how to connect them to the chip-level I/O.

3.6.5   Logic BIST

Logic BIST is a variation on the full scan approach. Where full scan must have its scan chain integrated into the chip's overall scan chain(s), logic BIST uses an LFSR (Linear Feedback Shift Register) to generate the test patterns locally. A signature recognition circuit checks the results of the scan test to verify correct behavior of the circuit.

Logic BIST has the advantage of keeping all pattern generation and checking within the macro. This provides some element of additional security against reverse engineering of the macro. It also reduces the requirements for scan memory in the tester and allows testing at higher clock rates than can be achieve on most testers. Logic BIST does require some additional design effort and some increase die area for the generator and checker, although tools to automate this process are becoming available.

Logic BIST is currently being used in some designs, but it is much less common than standard full-scan testing. The success of logic BIST in the long term probably depends on the ability of test equipment manufactures to keep up with the need for ever-increasing scan memory in the tester. If the test equipment fails to provide for scan test of large chips, logic BIST will become the test methodology of choice for SoC designs.

CHAPTER 4

The Macro Design Process

This chapter addresses the issues encountered in designing hard and soft macros for reuse. The topics include:

- An overview of the macro design workflow
- Contents of a design specification
- Top-level macro design and partitioning into subblocks
- Designing subblocks
- Integrating subblocks and macro verification
- Productization and prototyping issues

4.1 Design Process Overview

Once the SoC design team has developed a set of specifications for the various macros in the design, these macros need to be selected from an existing library of reusable parts or designed from scratch. This chapter describes the design process for developing macros, with an emphasis on developing reusable macros.

Figure 4-1 shows the macro design process up to the point of integrating subblocks back into the parent macro. Figure 4-2 shows the process of integrating the subblocks. The major steps in the macro design process are:

1. **Specification and partitioning** – The first thing the macro design team must do is to make sure it completely understands the initial macro specification. The team then refines the specification and partitions the design into subblocks. Usually this refinement includes developing a behavioral model and testing it. This is particularly useful in algorithmic-intensive designs, where the algorithm itself must be developed in addition to the implementation. It also provides an initial testbench and test suite for the macro, and it can be used to generate a simulation model for end users of the macro.

2. **Subblock specification and design** – Once the partitioning is complete, the designer develops a functional specification for the subblock, emphasizing the timing and functionality of the interfaces to other subblocks. The specifications for all subblocks are reviewed by the team and checked for consistency. The designer then develops the RTL code, the detailed timing constraints, the synthesis scripts, and the testbench and test suite for the subblock. Once these are completed and verified, the subblock is ready for integration with the other subblocks.

3. **Testbench development** – In parallel with the subblock development, some members of the macro design team refine the behavioral testbench into a testbench that can be used for RTL-level testing of the entire macro.

4. **Timing checks** – In addition, the team must be checking the timing budgets of the subblocks to ensure that they are consistent and achievable.

5. **Integration** – Integrating the subblocks into the macro includes generating the top-level netlist and using it to perform functional test and synthesis. The synthesis process includes verifying that the macro meets the requirements for manufacturing testability. This usually consists of doing scan insertion and automatic test pattern generation (ATPG) and verifying test coverage.

    Once these tasks have been successfully completed, the macro is ready for productization.

6. **Productization** – During the productization phase, the team prepares the macro for use by the SoC integration team. This process differs for soft macros and for hard macros. The differences are described later in this chapter.

Figure 4-3 shows the activities of the various team members during the first three phases of macro design.

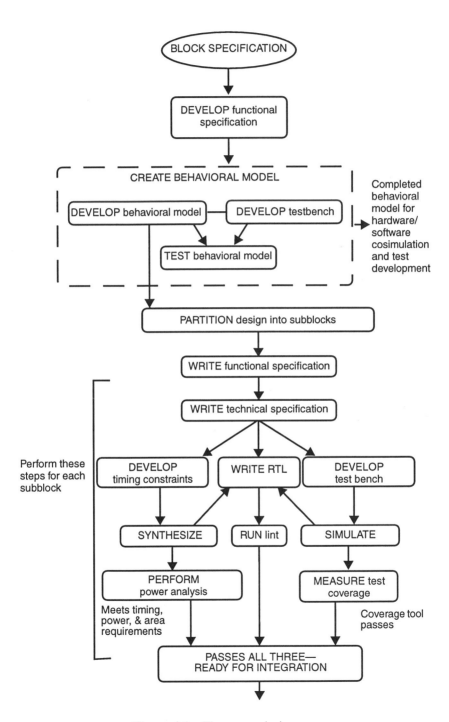

Figure 4-1   The macro design process

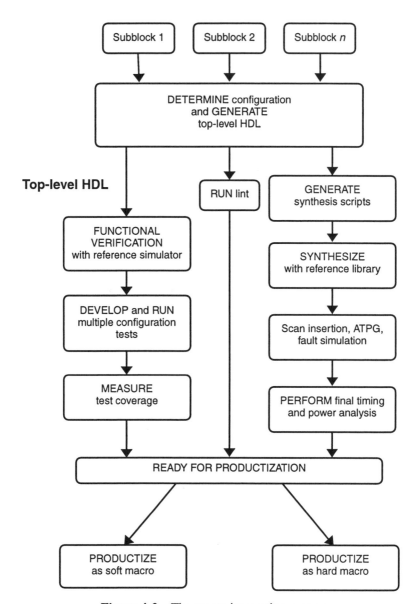

Figure 4-2   The macro integration process

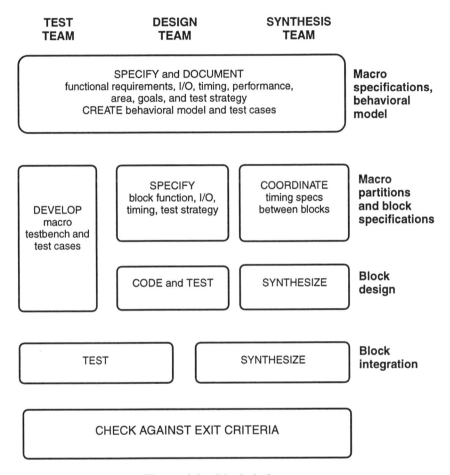

Figure 4-3   Block design teams

It is important to note that the separation of the design process into distinct phases does not imply a rigid, top-down design methodology. Frequently, some detailed design work must be done before the specification is complete, just to make sure that the design can be implemented.

A rigid, top-down methodology says that one phase cannot start until the preceding one is completed. We prefer a more mixed methodology, which simply says that one phase cannot complete until the preceding one is completed.

Methodology note – The design flow described in this chapter is the standard, RTL/synthesis flow. There are several alternate flows that use domain-specific tools such as Module Compiler. These flows are described in Chapter 6 of this manual.

4.2   Contents of a Design Specification

Specifications occur at every level in the design process. They are very general at the beginning and become progressively more focused and detailed as the design process continues. There are common elements, however, to all specifications. This section describes the archetypal structure of a good specification. When there are references to specifications later in this document, assume that the specifications contain the following elements:

Overview

This section briefly describes the technical goals for the design. In particular, if the design needs to comply with a specific standard, such as an IEEE standard, the standard must be specified here.

Functional requirements

This section describes the project from a technical perspective. Its main purpose is to describe the unit being designed as seen by the outside world: its form, fit, and function, and how it transforms the data at its inputs and outputs, based on the values of the software registers.

Physical requirements

This section describes the packaging, die size, power, and other physical design requirements of the unit being designed. For soft macros, it includes the cell libraries the design must support and the performance requirements for the design.

Design requirements

This section describes the design rules to which the design must comply. It may reference a standard design guideline document or explicitly list the guidelines. The issues addressed in this section of the specification are those described in Chapter 3 of this manual.

The block diagram

Block diagrams are essential for communicating the function of most hardware. The block diagrams must present enough detail so that the reader can understand the functionality of the unit being designed.

Interfaces to external system

This section describes the inputs and outputs of the module and how they operate:

- Signal names and functions
- Transaction protocols with cycle-accurate timing diagrams
- Legal values for input and output data
- Timing specifications
- Setup and hold times on inputs
- Clock to out times for outputs

- Special signals
- Asynchronous signals and their timing
- Clock, reset, and interrupt signals and their timing

Manufacturing test methodology

This section describes the manufacturing test methodology that the macro supports, and the chip-level requirements for supporting the test methodology. For most macros this methodology will be full scan. Typically, the integrator will perform scan insertion and ATPG on the entire chip (or a large section of the chip) at one time, rather than doing scan insertion for the macro and then integrating it into the chip design. Any untestable regions in the design must be specified.

For some hard macros, the performance penalty of scan-based testing is not acceptable, and parallel vectors are used for test. In this case, a JTAG-based boundary scan technique is used to isolate the macro and provide a way to apply the vectors to the block.

The software model

This section describes the hardware registers that are visible to the software. It includes complete information on which registers are read, write, and read/write, which bits are valid, and the detailed function of the register.

Software requirements

Hardware design doesn't stop until software runs on it. One of the key obligations of the hardware team is to provide the lowest level of software required to configure and operate the hardware. Once this software is provided, the software team only needs to know about these software routines, and not about the detailed behavior of the hardware or of the registers. For many hardware systems, this low-level software is referred to as the *set of software drivers for the system*. Although the drivers are often written by the software team, the hardware team is responsible for helping to specify this software and for verifying that it is correct.

The specification of this software must be included in the functional specification.

4.3   Top-Level Macro Design

The first phase of macro design consists of refining the functional specification to the point where the design can be partitioned into subblocks small enough that each subblock can be designed, coded, and tested by one person. The key to success in this phase is a complete and clear specification for the macro and its subblocks. In particular, the interfaces between subblocks must be clearly specified, so that subblock integration will be relatively smooth and painless.

4.3.1 Top-Level Macro Design Process

Figure 4-4 shows the top-level macro design process. This phase is complete when the design team has produced and reviewed the following top-level design elements:

- Updated macro hardware specification

  All sections of the document should be updated to reflect the design refinement that occurs during the macro top-level design process. In particular, the partitioning of the macro and the specifications for the subblocks must be added to the macro specification.

- Executable specification/behavioral model

  In many cases, a behavioral model is extremely useful as an executable specification for the macro. This model allows the development and debug of testbenches and test suites during the detailed design of the macro, rather than after the design is completed. For hard macros, this behavioral model can provide a key simulation model for the end user.

  A behavioral model is particularly useful for a macro that has a high algorithmic content. For a macro dominated by state machines and with little algorithmic content, a behavioral model may be of little use, because it would have all the interesting behavior abstracted out of it.

- Testbench

  A high-level, self-checking testbench with a complete set of test suites is essential to the successful design and deployment of the macro. Typically, the testbench consists of bus functional models for the surrounding system and is designed to allow the verification engineer to write tests at a relatively high level of abstraction.

- Preliminary specification for each subblock

4.3.2 Activities and Tools

The top-level macro design process involves the following activities and tools:

Develop algorithms and behavioral models

For most designs, the behavioral model is developed in C/C++, Verilog, or VHDL. C/C++ is particularly useful for designs that require significant hardware/software cosimulation, such as processor designs. Verilog and VHDL are preferred for designs in which some of the RTL characteristics, such as I/O behavior, may be needed.

C/C++, Verilog, and VHDL behavioral models are all easily ported to multiple environments. In particular, through the SWIFT interface, it is possible to package the model for secure, highly portable distribution to most commercial simulators.

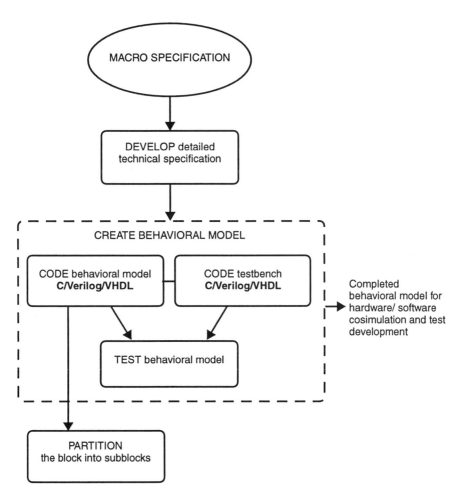

Figure 4-4   Flow for top-level macro design

Stream-driven tools such as COSSAP and SPW can be useful modeling tools
for those datapath-intensive designs in which the algorithm itself, independent
of the implementation, requires significant exploration and development. For
instance, when verifying a video compression algorithm, it may be necessary
to simulate with many frames of video. The different processing blocks in the
algorithm typically operate at different data rates; however, including the logic
to handle these different rates can slow down simulation. With a stream or data
driven simulator, each block executes as soon as the required data is received.
This approach provides the kind of simulation performance and ease of modeling required for datapath-intensive designs like video processing.

COSSAP can help generate RTL code and can assist in the hardware/software partitioning.

Develop testbenches

Testbench design and test development are essential and challenging at every level of representation—behavioral, RTL, and gate. For a full discussion of the macro testbench, refer to Chapter 7 of this manual.

4.4   Subblock Design

The second phase of macro design consists of design, RTL coding, and testing the subblocks in the macro. The key to the success of this phase is to have a complete and clear specification for each subblock before RTL coding begins, and to have a clear understanding of the deliverables needed at the end of the design phase.

4.4.1   Subblock Design Process

Subblock design, as illustrated in Figure 4-5, begins when there is a preliminary hardware specification for the subblock and a set of design guidelines for the project. The phase is complete when the design team has produced and reviewed the following subblock design elements:

- An updated hardware specification for the subblock
- A synthesis script
- A testbench for the subblock, and a verification suite that achieves 100% test coverage. See Chapter 7 for details on this requirement. In particular, note that the testbench/verification suite must provide 100% path and statement coverage as measured by a coverage tool.
- RTL that passes lint and synthesis. The final RTL code for the subblock must comply with the coding guidelines adopted by the design team. A configurable lint-like tool that verifies compliance to the guidelines is essential to ensure consistent code quality throughout the macro.

  The final RTL code must also synthesize on the target library and meet its timing constraints, using a realistic wire load model.

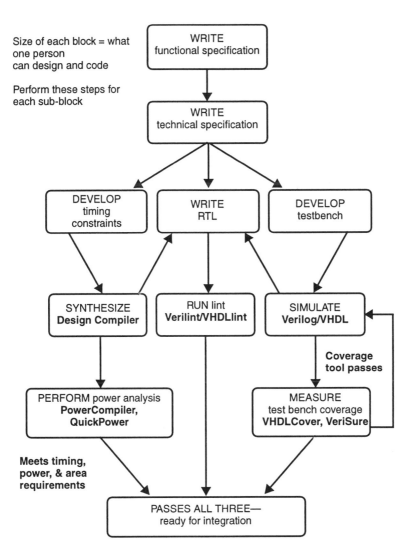

Figure 4-5   Flow for designing subblocks

4.4.2   Activities and Tools

The subblock design process involves the following activities and tools:

Develop the functional and technical specifications
The actual design of the subblock should be done before, not during, RTL coding.

The *functional specification* for the subblock describes, in detail, the aspects of the subblock that are visible to the rest of the macro: functionality, I/O, timing, area, and power. This specification can be included as part of the macro functional specification.

The *technical specification* describes the internals of the subblock and is intended to be the vehicle by which the designer captures the details of the subblock design before beginning coding. The quality of the technical specification is a key factor in determining the time required for the rest of the subblock design process. A good technical specification allows the designer to code once and to verify quickly. A poorly thought-out specification results in many iterations through the code/test/synthesis loop.

Develop RTL
In most cases, the RTL code is written directly by the designer.

For some arithmetic-intensive designs, Module Compiler provides a means of specifying the datapath and controlling the structures to be synthesized. Module Compiler generates a gate-level netlist and a simulation model for the subblock. It takes as input its own Verilog-like HDL. See "RAM and Datapath Generators" in Chapter 6 for a more detailed description of the work flow using Module Compiler.

Develop testbench
The design of the subblock-level testbench is described in Chapter 7. The critical requirements for this testbench are readability and ease of modification, so that the designer can easily create and extend the testbench, and use the testbench to detect and debug problems in the subblock.

Develop synthesis scripts and synthesize
The external timing constraints should be fully defined by the specification before coding begins. Synthesis scripts must be developed early in the design process and synthesis should begin as soon as the RTL code passes the most basic functional tests. These early synthesis runs give great insight into problem areas for timing and may significantly affect the final code.

Run lint
A lint-like tool, such as Verilint/VHDLlint from InterHDL, provides a powerful method for checking the RTL for violations of coding guidelines and other kinds of errors. It should be run often throughout the design process, since it is

the fastest means of catching errors. The final code must pass all lint checks specified in the coding guidelines.

Measure testbench coverage

It is essential to catch bugs as early as possible in the design process, since the time to find and correct a bug increases by an order of magnitude at each level of design integration. A bug found early during specification/behavioral modeling is dramatically cheaper than a bug found at macro integration.

Coverage tools such as VeriSure and VHDLCover provide a means of measuring statement and path coverage for RTL designs and testbenches. A coverage tool must be run on the final design and it should indicate 100 percent statement and path coverage before the subblock is integrated with other subblocks.

Perform power analysis

If power consumption is an issue, the design team uses QuickPower or Power Compiler to analyze power and to ensure that power consumption is within specification.

4.5   Macro Integration

The third phase of macro design consists of integrating the subblocks into the top-level macro and performing a final set of tests. The key to the success of this phase is to have subblocks that have been designed to the guidelines outlined in this document. In particular, the timing and functional behavior of the interfaces between subblocks should be completely specified before subblock design and verified after subblock design. Most bugs occur at the interfaces between subblocks and as a result of misunderstandings between members of the design team.

4.5.1   Integration Process

The macro integration process, shown in Figure 4-6, is complete when:

- Development of top-level RTL, synthesis scripts, and testbenches is complete
- Macro RTL passes all tests
- Macro synthesizes with reference library and meets all timing, area, and power criteria
- Macro RTL passes lint and manufacturing test coverage

The only new criterion here is the one about meeting the manufacturing test coverage requirements. Most macros use a full scan methodology for manufacturing test, and require 95 percent coverage (99 percent is preferred). Whatever the methodology, test coverage must be measured at this point and must be proven to meet the requirements in the functional specification.

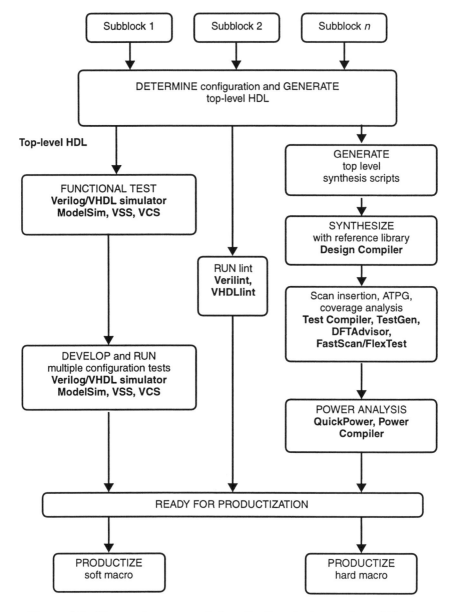

Figure 4-6   Flow and recommended applications for integrating subblocks

4.5.2   Activities and Tools

The process of integrating the subblocks into the top-level macro involves the following activities and tools:

Develop top-level RTL

Once the subblocks have all been developed, the design team needs to develop a top-level RTL description that instantiates the subblocks and connects them together. Parameterizable macros, where the number of instances of a particular subblock may vary, present a particular challenge here. It may be necessary to develop a script that will generate the appropriate instances and instantiate them in the top level RTL.

Run functional tests

It is essential to develop a thorough functional test suite and to run it on the final macro design. The design team must run this test on a sufficient set of configurations to ensure that the macro is robust for all possible configurations.

The verification strategy for the entire macro is discussed in Chapter 7 of this manual.

Develop synthesis scripts

Once the subblock-level synthesis scripts have all been developed, the design team needs to develop a top-level synthesis script. For parameterizable macros, where the number of instances of a particular subblock may vary, this presents a particular challenge. It may be necessary to provide a set of scripts for different configurations of the macro. It may also be useful to provide different scripts for different synthesis goals: one script to achieve optimal timing performance, another to minimize area.

Run synthesis

The design team must run synthesis on a sufficiently large set of configurations to ensure that synthesis will run successfully for all configurations. In general, this means synthesizing both a minimum and maximum configuration. Note that the final synthesis constraints must take into account the fact that scan will later be inserted in the macro, adding some setup time requirements to the flops.

Use Design Compiler to perform top-level synthesis.

Perform scan insertion

The final RTL code must also meet the testability requirements for the macro. Most macros will use a full scan test methodology and require 95% coverage (99% preferred).

Use a test insertion tool (for example Test Compiler, TestGen, DFTAdvisor, or FastScan/FlexTest) to perform scan insertion and automatic test pattern generation for the macro. As part of this process, the test insertion tool should also report the actual test coverage for the macro.

After scan insertion, the design team uses a static timing analysis tool to verify the final timing of the macro.

Perform power analysis

If power consumption is an issue, the design team uses QuickPower or Power Compiler to analyze power and to ensure that power consumption is within specification.

Run lint

Finally, run the lint tool on the entire design to ensure compliance to guidelines. In addition, use the lint tool to verify the translatability of the macro and testbench between Verilog and VHDL.

4.6   Soft Macro Productization

The final phase of macro design consists of productizing the macro, which means creating the remaining deliverables that system integrators will require for reuse of the macro. This chapter describes the productization of soft macros only. The development and productization of hard macros is described in Chapter 8.

4.6.1   Productization Process

The soft macro productization phase, shown in Figure 4-7, is complete when the design team has produced and reviewed the following components of the final product.

- Verilog and VHDL versions of the code, testbenches, and tests
- Supporting scripts for the design

  This includes the installation scripts and synthesis scripts required to build the different configurations of the macro.

- Documentation

  This includes updating all the functional specifications and generating the final user documentation from them.

- Final version locked in RCS

  All deliverables must be in a revision control system to allow future maintenance.

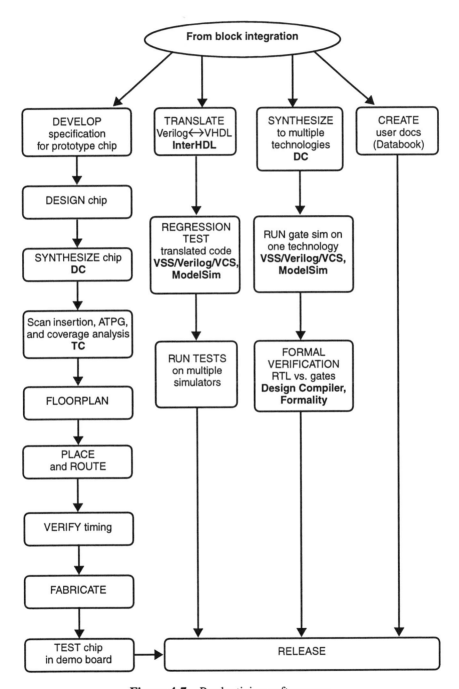

Figure 4-7   Productizing soft macros

4.6.2   Activities and Tools

The soft macro productization process involves the following activities and tools:

Develop a prototype chip

A prototype chip is essential for verifying both the robustness of the design and the correctness of the original specifications. Some observers estimate that 90 percent of chips work the first time but only 50 percent of chips work correctly in the system.

Developing a chip using the macro and testing it in a real application with real application software allows us to:

- Verify that the design is functionally correct.
- Verify that the design complies with the appropriate standards (for instance, we can take a PCI test chip to the PCI SIG for compliance testing).
- Verify that the design is compatible with the kind of hardware/software environment that other integrators are likely to use.

The process for developing the prototype chip is a simple ASIC flow appropriate for small chip design. It is assumed that the chip will be a simple application of the macro, perhaps twice the size of the macro itself in gate count.

Provide macro and testbench in both Verilog and VHDL

To be widely useful, the macro and its testbenches must be available in both the Verilog and VHDL languages. Commercial translators are available, including one from InterHDL. These translators do a reasonable job on RTL code but still present some challenge for translating testbenches.

After the code and testbenches have been translated, they must be re-verified to validate the translation.

Test on several simulators

In addition, the macro and testbenches should be run on the most popular simulators in order to ensure portability. This is particularly important for the VHDL simulators, which have significant differences from vendor to vendor.

Synthesize on multiple technologies

The macro should be synthesized using a variety of technologies to ensure portability of the scripts and to ensure that the design can meet its timing and area goals with the ASIC libraries that customers are most likely to use.

Perform gate-level simulation

Gate-level simulation must be run on at least one target technology in order to verify the synthesis scripts.

Formal verification

Using formal verification tools, such as Formality or the `compare_design` feature of Design Compiler, we can verify that the final netlist is functionally equivalent to the original RTL.

Create/update user documentation

The functional specifications created during the design process are usually not the best vehicle for helping a customer use the macro. A set of user documents must be developed that address this need. The components of this documentation are described in Chapter 9 of this manual.

CHAPTER 5

RTL Coding Guidelines

This chapter offers a collection of coding rules and guidelines. Following these practices helps to ensure that your HDL code is readable, modifiable, and reusable. Following these coding practices also helps to achieve optimal results in synthesis and simulation.

Topics in this chapter include:

- Basic coding practices
- Coding for portability
- Guidelines for clocks and resets
- Coding for synthesis
- Partitioning for synthesis
- Designing with memories
- Code profiling

5.1 Overview of the Coding Guidelines

The coding guidelines in this chapter are based on a few fundamental principles. The basic underlying goal is to develop RTL code that is simple and regular. Simple and regular structures are inherently easier to design, code, verify, and synthesize than more complex designs. The overall goal for any reusable design should be to keep it as simple as possible and still meet its functional and performance goals.

The coding guidelines detailed in this chapter provide the following general recom-
mendations:

- Use simple constructs, basic types (for VHDL), and simple clocking schemes.
- Use a consistent coding style, consistent naming conventions, and a consistent
  structure for processes and state machines.
- Use a regular partitioning scheme, with all module outputs registered and with
  modules roughly of the same size.
- Make the RTL code easy to understand, by using comments, meaningful names,
  and constants or parameters instead of hard-coded numbers.

By following these guidelines, the developer should be better able to produce code
that converges quickly to the desired performance, in terms of functionality, timing,
power, and area.

5.2   Basic Coding Practices

The following guidelines address basic coding practices, focusing on lexical conventions and basic RTL constructs.

5.2.1   General Naming Conventions

Rule – Develop a naming convention for the design. Document it and use it consistently throughout the design.

Guideline – Use lowercase letters for all signal names, variable names, and port
names.

Guideline – Use uppercase letters for names of constants and user-defined types.

Guideline – Use meaningful names for signals, ports, functions, and parameters. For
example, do not use *ra* for a RAM address bus. Instead, use *ram_addr*.

Guideline – If your design uses several parameters, use short but descriptive names.
During elaboration, the synthesis tool concatenates the module's name, parameter
names, and parameter values to form the design unit name. Thus, lengthy parameter
names can cause excessively long design unit names when you elaborate the design
with Design Compiler.

Guideline – Use the name *clk* for the clock signal. If there is more than one clock in
the design, use *clk* as the prefix for all clock signals (for example, *clk1*, *clk2*, or
clk_interface).

Guideline – Use the same name for all clock signals that are driven from the same source.

Guideline – For active low signals, end the signal name with an underscore followed by a lowercase character (for example, _b or _n). Use the same lowercase character to indicate active low signals throughout the design.

Guideline – For standardization, we recommend that you use _n to indicate an active low signal. However, any lowercase character is acceptable as long as it is used consistently.

Guideline – Use the name *rst* for reset signals. If the reset signal is active low, use *rst_n* (or substitute *n* with whatever lowercase character you are using to indicate active low signals).

Rule – When describing multibit buses, use a consistent ordering of bits. For VHDL, use either (y downto x) or (x to y). For Verilog, use (x:0) or (0:x). Using a consistent ordering helps improve the readability of the code and reduces the chance of accidently swapping order between connected buses.

Guideline – Although the choice is somewhat arbitrary, we recommend using (y downto x) for multibit signals in VHDL and (x:0) for multibit signals in Verilog. We make this recommendation primarily to establish a standard, and thus achieve some consistency across multiple designs and design teams.
See Example 5-1.

Example 5-1 Using downto in port declarations

```
entity DW_addinc is
  generic(WIDTH : natural);
  port(
    A,B : in std_logic_vector(WIDTH-1 downto 0);
    CI  : in std_logic;
    SUM : out std_logic_vector(WIDTH-1 downto 0);
    CO  : out std_logic;
  );
end DW01_addinc;
```

Guideline – When possible, use the same name or similar names, for ports and signals that are connected (for example, a => a; or a => a_int;).

Guideline – When possible, use the signal naming conventions listed in Table 5-1.

Table 5-1   Signal naming conventions

Convention	Use
*_r	Output of a register (for example, `count_r`)
*_a	Asynchronous signal (for example, `addr_strobe_a`)
*_p*n*	Signal used in the *n*th phase (for example, `enable_p2`)
*_nxt	Data before being registered into a register with the same name
*_z	Three-state internal signal

5.2.2   Naming Conventions for VITAL Support

VITAL is a gate-level modeling standard for VHDL libraries and is described in IEEE Specification 1076.4. This specification places restrictions on the naming conventions (and other characteristics) of the port declarations at the top level of a library element.

Normally, an RTL coding style document need not address gate-level modeling conventions. However, some of these issues can affect developers of hard macros. The deliverables for a hard macro include full-functional/full-timing models, where a timing wrapper is added to the RTL code. If the timing wrapper is in VHDL, then it must be VITAL-compliant.

Background

According to IEEE Specification 1076.4, VITAL libraries can have two levels of compliance with the standard: VITAL_Level0 and VITAL_Level1. VITAL_Level1 is more rigorous and deals with the architecture (functionality and timing) of a library cell. VITAL_Level0 is the interface specification that deals with the ports and generics specifications in the entity section of a VHDL library cell. VITAL_Level0 has strict rules regarding naming conventions and port/generic types. These rules were designed so that simulator vendors can assume certain conventions and deal with SDF back-annotation in a uniform manner.

Rules

Section 4.3.1 of IEEE Specification 1076.4 addresses port naming conventions and includes the following rules. These restrictions apply only to the top-level entity of a hard macro.

Rule (hard macro, top-level ports) – Do not use underscore characters (_) in the entity port declaration for the top-level entity of a hard macro.

The reason for the above rule is that VITAL uses underscores as separators to construct names for SDF back-annotation from the SDF entries.

Rule (hard macro, top-level ports) – A port that is declared in entity port declaration shall not be of mode LINKAGE.

Rule (hard macro, top-level ports) – The type mark in an entity port declaration shall denote a type or subtype that is declared in package `std_logic_1164`. The type mark in the declaration of a scalar port shall denote a subtype of `std_ulogic`. The type mark in the declaration of an array port shall denote the type `std_logic_vector`.

Rule (hard macro, top-level ports) – The port in an entity port declaration cannot be a guarded port. Furthermore, the declaration cannot impose a range constraint on the port, nor can it alter the resolution of the port from that defined in the standard logic package.

5.2.3   Architecture Naming Conventions

Guideline – Use the VHDL architecture types listed in Table 5-2.

Table 5-2   Architecture naming conventions

Architecture	Naming Convention

synthesis model	`ARCHITECTURE rtl OF my_syn_model IS`
	or
	`ARCHITECTURE str OF my_structural_design IS`

simulation model	`ARCHITECTURE sim OF my_behave_model IS`
	or
	`ARCHITECTURE tb OF my_test_bench IS`

5.2.4   Include Headers in Source Files

Rule – Include a header at the top of every source file, including scripts. The header must contain:

- Filename
- Author

- Description of function and list of key features of the module
- Date the file was created
- Modification history including date, name of modifier, and description of the change

Example 5-2 shows a sample HDL source file header.

Example 5-2   Header in an HDL source file

```
--This confidential and proprietary software may be used
--only as authorized by a licensing agreement from
--Synopsys Inc.
--In the event of publication, the following notice is
--applicable:
--
-- (C) COPYRIGHT 1996 SYNOPSYS INC.
-- ALL RIGHTS RESERVED
--
-- The entire notice above must be reproduced on all
--authorized copies.
--
-- File       : DWpci_core.vhd
-- Author     : Jeff Hackett
-- Date       : 09/17/96
-- Version    : 0.1
-- Abstract   : This file has the entity, architecture
--               and configuration of the PCI 2.1
--               MacroCell core module.
--               The core module has the interface,
--               config, initiator,
--               and target top-level modules.
--
-- Modification History:
-- Date        By     Version    Change Description
--
============================================================
-- 9/17/96    JDH      0.1        Original
-- 11/13/96   JDH                 Last pre-Atria changes
-- 03/04/97   SKC                 changes for ism_ad_en_ffd_n
--                                and tsm_data_ffd_n
--
============================================================
```

5.2.5   Use Comments

Rule – Use comments appropriately to explain all processes, functions, and declarations of types and subtypes. See Example 5-3.

Example 5-3   Comments for a subtype declaration

```
--Create subtype INTEGER_256 for built-in error
--checking of legal values.
subtype INTEGER_256 is type integer range 0 to 255;
```

Guideline – Use comments to explain ports, signals, and variables, or groups of signals or variables.

Comments should be placed logically, near the code that they describe. Comments should be brief, concise, and explanatory. Avoid "comment clutter"; obvious functionality does not need to be commented. The key is to describe the intent behind the section of code.

5.2.6   Keep Commands on Separate Lines

Rule – Use a separate line for each HDL statement. Although both VHDL and Verilog allow more than one statement per line, the code is more readable and maintainable if each statement or command is on a separate line.

5.2.7   Line Length

Guideline – Keep the line length to 72 characters or less.

Lines that exceed 80 characters are difficult to read in print and on standard terminal width computer screens. The 72 character limit provides a margin that enhances the readability of the code.

For HDL code (VHDL or Verilog), use carriage returns to divide lines that exceed 72 characters and indent the next line to show that it is a continuation of the previous line. See Example 5-4.

Example 5-4   Continuing a line of HDL code

```
hp_req <= (x0_hp_req or t0_hp_req or x1_hp_req or
    t1_hp_req or s0_hp_req or t2_hp_req or s1_hp_req or
    x2_hp_req or x3_hp_req or x4_hp_req or x5_hp_req or
    wd_hp_req and ea and pf_req nor iip2);
```

5.2.8   Indentation

Rule – Use indentation to improve the readability of continued code lines and nested loops. See Example 5-5.

Guideline – Use indentation of 2 spaces. Larger indentation (for example, 8 spaces) restricts line length when there are several levels of nesting.

Guideline – Avoid using tabs. Differences in editors and user setups make the positioning of tabs unpredictable and can corrupt the intended indentation. There are programs available, including language-specific versions of emacs, that will replace tabs with spaces.

Example 5-5   Indentation in a nested `if` loop

```
if (bit_width(m+1) >= 2) then
  for i in 2 to bit_width(m+1) loop
    spin_j := 0;
    for j in 1 to m loop
      if j > spin_j then
        if (matrix(m)(i-1)(j) /= wht) then
          if (j=m) and (matrix(m)(i)(j) = wht) then
            matrix(m)(i)(j) := j;
          else
            for k in j+1 to m loop
              if (matrix(m)(i-1)(k) /= wht) then
                matrix(m)(i)(k) := j;
                spin_j := k;
                exit;
              end if;
            end loop;  -- k
          end if;
        end if;
      end if;
    end loop;  -- j
  end loop;  -- i
end if;
```

5.2.9 Do Not Use HDL Reserved Words

Rule – Do not use VHDL or Verilog reserved words for names of any elements in your RTL source files. Because macro designs must be translatable from VHDL to Verilog and from Verilog to VHDL, it is important not to use VHDL reserved words in Verilog code, and not to use Verilog reserved words in VHDL code.

5.2.10 Port Ordering

Rule – Declare ports in a logical order, and keep this order consistent throughout the design.

Guideline – Declare one port per line, with a comment following it (preferably on the same line).

Guideline – Declare the ports in the following order:

Inputs:
- Clocks
- Resets
- Enables
- Other control signals
- Data and address lines

Outputs:
- Clocks
- Resets
- Enables
- Other control signals
- Data

Guideline – Use comments to describe groups of ports. See Example 5-6.

Example 5-6   Port ordering in entity definition

```
entity my_fir is
  generic  (
    DATA_WIDTH  : positive;
    COEF_WIDTH  : positive;
    ACC_WIDTH   : positive;
    ORDER       : positive
  );
```

```
  port (
--
--            Control Inputs
--
    clk      : in std_logic;
    rst_n    : in std_logic;
    run      : in std_logic;
    load     : in std_logic;
    tc       : in stᵈ logic;
--
--            Data Inɩ ɩts
--
    data_in : in std_logic_vector(DATA_WIDTH-1 downto
      0);
--
    coef_in : in std_logic_vector(COEF_WIDTH-1 downto
      0);
    sum_in  : in std_logic_vector(ACC_WIDTH-1  downto
      0);
--
--            Control Outputs
--
    start   : out std_logic;
    hold    : out std_logic;
--
--            Data Outputs
--

    sum_out : out std_logic_vector(ACC_WIDTH-1 downto
      0));

end my_fir;
```

5.2.11   Port Maps and Generic Maps

Rule – Always use explicit mapping for ports and generics, using named association rather than positional association. See Example 5-7.

Guideline – Leave a blank line between the input and output ports to improve readability.

Example 5-7 Using named association for port mapping

VHDL:

```
-- instantiate my_add
  U_ADD: my_add
    generic map (width => WORDLENGTH)
    port map (
      a => in1,
      b => in2,
      ci => carry_in,

      sum => sum,
      co => carry_out
    );
```

Verilog:

```
// instantiate my_add
my_add #('WORDLENGTH) U_ADD (
  .a    (in1     ),
  .b    (in2     ),
  .ci   (carry_in ),

  .sum  (sum     ),
  .co   (carry_out)
);
```

5.2.12 VHDL Entity, Architecture, and Configuration Sections

Guideline – Place entity, architecture, and configuration sections of your VHDL design in the same file. Putting all the information about a particular design in one file makes the design easier to understand and to maintain.

If you include subdesign configurations in a source file with entity and architecture declarations, you must comment them out for synthesis. You can do this with the `pragma translate_off` and `pragma translate_on` pseudo-comments in the VHDL source file, as shown in Example 5-8.

Example 5-8 Using pragmas to comment out VHDL configurations for synthesis

```
-- pragma translate_off
configuration cfg_example_struc of example is
  for struc
    use example_gate;
```

```
    end for;
  end cfg_example_struc;
  -- pragma translate_on
```

5.2.13 Use Functions

Guideline – Use functions when possible, instead of repeating the same sections of code. If possible, generalize the function to make it reusable. Also, use comments to explain the function.

For example, if your code frequently converts address data from one format to another, use a function to perform the conversion and call the function whenever you need to. See Example 5-9.

Example 5-9   Creating a reusable function

VHDL:

```
--This function converts the incoming address to the
--corresponding relative address.

function convert_address
    (input_address, offset : integer)
  return integer is
begin

  -- ... function body here ...

end; -- convert_address
```

Verilog:

```
// This function converts the incoming address to the
// corresponding relative address.

function ['BUS_WIDTH-1:0] convert_address;
  input input_address, offset;
  integer input_address, offset;

begin
  // ... function body goes here ...
end
endfunction // convert_address
```

5.2.14 Use Loops and Arrays

Guideline – Use loops and arrays for improved readability of the source code. For example, describing a shift register, PN-sequence generator, or Johnson counter with a loop construct can greatly reduce the number of lines of source code while still retaining excellent readability. See Example 5-10.

Example 5-10   Using loops to improve readability

```
shift_delay_loop: for i in 1 to (number_taps-1) loop
  delay(i) := delay(i-1);
end loop shift_delay_loop;
```

The ARRAY construct also reduces the number of statements necessary to describe the function and improves readability. The following is an example of a register bank implemented as a two-dimensional array of flip-flops. See Example 5-11.

Example 5-11   Register bank using an array

```
type reg_array is array(natural range <>) of
  std_logic_vector(REG_WIDTH-1 downto 0);
signal reg: reg_array(WORD_COUNT-1 downto 0);

begin
  REG_PROC: process(clk)
  begin
    if clk='1' and clk'event then
      if we='1' then
        reg(addr) <= data;
      end if;
    end if;
  end process REG_PROC;

  data_out <= reg(addr);
```

Guideline – Arrays are significantly faster to simulate than for loops. To improve simulation performance, use vector operations on arrays rather than for loops whenever possible. See Example 5-12.

Example 5-12   Using arrays for faster simulation

Poor coding style;

```
function my_xor( bbit : std_logic;
                 avec : std_logic_vector(x downto y) )
  return std_logic_vector is
variable cvec :
  std_logic_vector(avec'range-1 downto 0);
begin
  for i in avec'range loop        -- bit-level for loop
    cvec(i) := avec(i) xor bbit; -- bit-level xor
  end loop;
  return(cvec);
end;
```

Recommended coding style:

```
function my_xor( bbit : std_logic;
                 avec : std_logic_vector(x downto y) )
  return std_logic_vector is
variable cvec, temp :
  std_logic_vector(avec'range-1 downto 0);
begin
  temp := (others => bbit);
  cvec := avec xor temp;
  return(cvec);
end;
```

5.2.15   Use Meaningful Labels

Rule – Label each process block with a meaningful name. This is very helpful for debug. For example, you can set a breakpoint by referencing the process label.

Guideline – Label each process block *<name>*_PROC.

Rule – Label each instance with a meaningful name.

Guideline – Label each instance U_*<name>*.

In a multi-layered design hierarchy, keep the labels short as well as meaningful. Long process and instance labels can cause excessively long path names in the design hierarchy. See Example 5-13.

Rule – Do not duplicate any signal, variable, or entity names. For example, if you have a signal named *incr*, do not use *incr* as a process label.

Example 5-13   Meaningful process label

```
-- Synchronize requests (hold for one clock).
SYNC_PROC : process (req1, req2, rst, clk)

... process body here ...

end process SYNC_PROC;
```

5.3   Coding for Portability

The following guidelines address portability issues. By following these guidelines, you will create code that is technology-independent, compatible with various simulation tools, and easily translatable from VHDL to Verilog (or from Verilog to VHDL).

5.3.1   Use Only IEEE Standard Types

Rule (VHDL only) – Use only IEEE standard types.

You can create additional types and subtypes, but all types and subtypes should be based on IEEE standard types. Example 5-14 shows how to create a subtype (`word_type`) based on the IEEE standard type `std_logic_vector`.

Example 5-14   Creating a subtype from `std_logic_vector`

```
--Create new 16-bit subtype
subtype WORD_TYPE is std_logic_vector (15 downto 0);
```

Guideline (VHDL only) – Use `std_logic` rather than `std_ulogic`. Likewise, use `std_logic_vector` rather than `std_ulogic_vector`. The `std_logic` and `std_logic_vector` types provide the resolution functions required for three-state buses. The `std_ulogic` and `std_ulogic_vector` types do not provide resolution functions.

Note – Standardizing on either `std_logic` or `std_ulogic` is more important than which of the two you select. There are advantages and disadvantages to each. Most designers today use `std_logic`, which is somewhat better supported by EDA tools. In most applications, the availability of resolution functions is not required. Internal three-state buses present serious design challenges and should be used only when absolutely necessary. However, at the system level and in those extreme cases where internal three-state buses required, the resolution functions are essential.

Guideline (VHDL only) – Be conservative in the number of subtypes you create. Using too many subtypes makes the code difficult to understand.

Guideline (VHDL only) – Do not use the types bit or bit_vector. Many simulators do not provide built-in arithmetic functions for these types. Example 5-15 shows how to use built-in arithmetic packages for std_logic_vector.

Example 5-15   Using built-in arithmetic functions for std_logic_vector

```
use ieee.std_logic_arith.all;
signal a,b,c,d:std_logic_vector(y downto x);
    c <= a + b;
```

5.3.2   Do Not Use Hard-Coded Numeric Values

Guideline – Do not use hard-coded numeric values in your design. As an exception, you can use the values 0 and 1 (but not in combination, as in 1001). Example 5-16 shows Verilog code that uses a hard-coded numerical value (7) in the "poor coding style" example and a constant (MY_BUS_SIZE) in the "recommended coding style" example.

Example 5-16   Using constants instead of hard-coded values

Poor coding style:

```
wire     [7:0] my_in_bus;
reg      [7:0] my_out_bus;
```

Recommended coding style:

```
`define MY_BUS_SIZE 8
wire     [`MY_BUS_SIZE-1:0] my_in_bus;
reg      [`MY_BUS_SIZE-1:0] my_out_bus;
```

5.3.3   Packages

Guideline (VHDL only) – Collect all parameter values and function definitions for a design into a single separate file (a "package") and name the file *DesignName*_package.vhd.

5.3.4 Include Files

Guideline (Verilog only) – Keep the `define` statements for a design in a single separate file and name the file *DesignName*_params.v.

5.3.5 Avoid Embedding dc_shell Scripts

Although it is possible to embed dc_shell synthesis commands directly in the source code, this practice is not recommended. Others who synthesize the code may not be aware of the hidden commands, which may cause their synthesis scripts to produce poor results. If the design is reused in a new application, the synthesis goals may be different, such as a higher-speed version. If the source code is reused with a new release of Design Compiler, the commands will still be supported but may be obsolete.

There are several exceptions to this rule. In particular, the synthesis directives to turn synthesis on and off must be embedded in the code in the appropriate places. These exceptions are noted in various guidelines throughout this chapter.

5.3.6 Use Technology-Independent Libraries

Guideline – Use the DesignWare Foundation Library to maintain technology independence.

The DesignWare Foundation Library contains improved architectures for the inferable arithmetic components, such as:

- Adders
- Multipliers
- Comparators
- Incrementers and decrementers

These architectures provide improved timing performance over the equivalent internal Design Compiler architectures.

The DesignWare Foundation Library also provides additional arithmetic components such as:

- Sin, cos
- Modulus, divide
- Square root
- Arithmetic and barrel shifters

These DesignWare components are all high-performance designs that are portable across processes. They provide significantly more portability than custom-designed, process-specific designs. Using these components helps you create designs that achieve high performance in all target libraries.

The DesignWare Foundation Library also includes a number of sequential components, also designed to be completely process-portable, and which can save considerable design time. These components include:

* FIFO's and FIFO controllers
* ECC
* CRC
* JTAG components and ASIC debugger

For more information about using DesignWare components, see the *DesignWare Foundation Library Databook* and the *DesignWare User Guide*.

Guideline – Avoid instantiating gates in the design. Gate-level designs are very hard to read, and thus difficult to maintain and reuse. If technology-specific gates are used, then the design is not portable to other technologies.

Guideline – If you must use technology-specific gates, then isolate these gates in a separate module. This will make it easier to modify these gates as needed for different technologies.

Guideline – The GTECH library If you must instantiate a gate, use a technology-independent library such as the Synopsys generic technology library, GTECH. This library contains the following technology-independent logical components:

* AND, OR, and NOR gates (2, 3, 4, 5, and 8)
* 1-bit adders and half adders
* 2-of-3 majority
* Multiplexors
* Flip-flops
* Latches
* Multiple-level logic gates, such as AND-NOT, AND-OR, AND-OR-INVERT

5.3.7   Coding For Translation (VHDL to Verilog)

Guideline (VHDL only) – Do not use `generate` statements. There is no equivalent construct in Verilog.

Guideline (VHDL only) – Do not use `block` constructs. There is no equivalent construct in Verilog.

Guideline (VHDL only) – Do not use code to modify `constant` declarations. There is no equivalent capability in Verilog.

5.4   Guidelines for Clocks and Resets

The following section contains guidelines for clock and reset signals. The basic theory behind these guidelines is that a simple clocking structure is easier to understand, analyze, and maintain. It also consistently produces better synthesis results. The preferred clocking structure is a single global clock and positive edge-triggered flops as the only sequential devices, as illustrated in Figure 5-1.

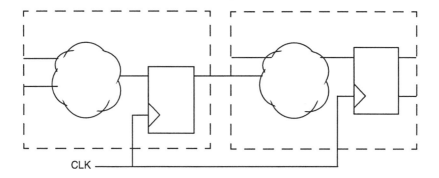

Figure 5-1   Ideal Clocking Structure

5.4.1   Avoid Mixed Clock Edges

Guideline – Avoid using both positive-edge and negative-edge triggered flip-flops in your design.

Mixed clock edges may be necessary in some designs. In designs with very aggressive timing goals, for example, it may be necessary to capture data on both edges of the clock. However, clocking on both edges creates several problems, and should be used with caution:

- The duty cycle of the clock becomes a critical issue in timing analysis, in addition to the clock frequency itself.
- Most scan-based testing methodologies require separate handling of positive and negative-edge triggered flops.

Figure 5-2 shows an example of a module with both positive-edge and negative-edge triggered flip-flops.

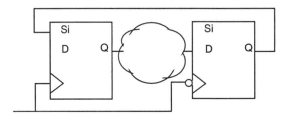

Figure 5-2   Bad example: Mixed clock edges

Rule – If you must use both positive-edge and negative-edge triggered flip-flops in your design, be sure to model the worst case duty cycle of the clock accurately in synthesis and timing analysis.

The assumption of a perfect clock with 50% duty cycle is optimistic, giving signals half the clock cycle to propagate from one register to the next. In the physical design, the duty cycle will be not be perfect, and the actual time available for signals to propagate can be much smaller.

Rule – If you must use both positive-edge and negative-edge triggered flip-flops in your design, be sure to document the assumed duty cycle in the user documentation.

In most chip designs, the duty cycle is a function of the clock tree that is inserted into the design; this clock tree insertion is usually specific to the process technology. The chip designer using the macro must check that the actual duty cycle will match requirements of the macro, and must know how to change the synthesis/timing analysis scripts for the macro to match the actual conditions.

Guideline – If you must use a large number of both positive-edge and negative-edge triggered flip-flops in your design, it may be useful to separate them into different modules. This makes it easier to identify the negative-edge flops, and thus to put them in different scan chains.

Figure 5-3 shows an example design where the positive-edge triggered flip-flops and negative-edge triggered flip-flops are partitioned into separate blocks.

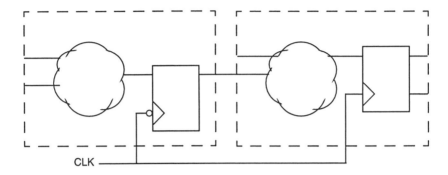

Figure 5-3   Better example: Negative-edge and positive-edge flip-flops are separated

5.4.2   Avoid Clock Buffers

Guideline – Avoid hand instantiating clock buffers in RTL code. Clock buffers are normally inserted after synthesis as part of the physical design. In synthesizable RTL code, clock nets are normally considered ideal nets, with no delays. During place and route, the clock tree insertion tool inserts the appropriate structure for creating as close to an ideal, balanced clock distribution network as possible.

5.4.3   Avoid Gated Clocks

Guideline – Avoid gated clocks in your design. Clock gating circuits tend to be technology specific and timing dependent. Improper timing of a gated clock can generate a false clock or glitch, causing a flip-flop to clock in the wrong data. Also, the skew of different local clocks can cause hold time violations.

Gated clocks also cause limited testability because the logic clocked by a gated clock cannot be made part of a scan chain. Figure 5-4 shows a design where U2 cannot be clocked during scan-in, test, or scan-out, and cannot be made part of the scan chain.

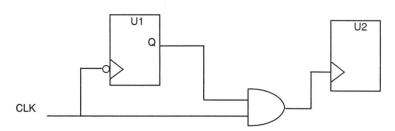

Figure 5-4   Bad example: Limited testability and skew problems because of gated
clock

5.4.4   Avoid Internally Generated Clocks

Guideline – Avoid using internally generated clocks in your design.

Internally generated clocks cause limited testability because logic that they clock cannot be made part of a scan chain. Internally generated clocks also make it more difficult to constrain the design for synthesis.

Figure 5-5 shows a design in which U2 cannot be clocked during scan-in, test, or
scan-out, and cannot be made part of the scan chain because it is clocked by an internally generated clock. As an alternative, design synchronously or use multiple clocks.

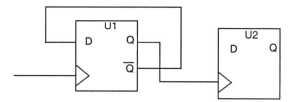

Figure 5-5   Bad example: Internally generated clock

5.4.5   Gated Clocks and Low Power Designs

Some designs, especially low-power designs, required a gated clocks. The following
guidelines address this issue.

Guideline – If you must use a gated clock, or an internally generated clock or reset,
keep the clock and/or reset generation circuitry as a separate module at the top level
of the design. Partition the design so that all the logic in a single module uses a single
clock and a single reset. See Figure 5-6.

In particular, a gated clock should never occur within a macro. The clock gating circuit, if required, should appear at the top level of the design hierarchy, as shown in Figure 5-6.

Isolating clock and reset generation logic in a separate module solves a number of problems. It allows submodules 1-3 to use the standard timing analysis and scan insertion techniques. It restricts exceptions to the RTL coding guidelines to a small module than can be carefully reviewed for correct behavior. It also makes it easier for the design team to develop specific test strategies for the clock/reset generation logic.

Guideline – If your design requires a gated clock, model it using synchronous load registers, as illustrated in Example 5-17.

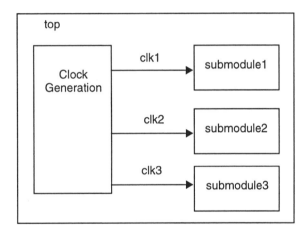

Figure 5-6   Good example: Clock generation circuitry is isolated at the top level

Example 5-17   Use synchronous load instead of combinational gating

Poor coding style:

```
clk_p1 <= clk and p1_gate;
EX17A_PROC: process (clk_p1)
  begin
    if (clk_p1'event and clk_p1 = '1') then
        . . . . . . .
    end if;
  end process EX17A_PROC;
```

Good coding style:

```
EX17B_PROC: process (clk)
  begin
    if (clk'event and clk = '1') then
      if (p1_gate = '1') then
          . . .
      end if;
    end if;
  end process EX17B_PROC;
```

5.4.6   Avoid Internally Generated Resets

Make sure your registers are controlled only by a simple reset signal.

Guideline – Avoid internally generated, conditional resets if possible. Generally, all the registers in the macro should be reset at the same time. This approach makes analysis and design much simpler and easier.

Guideline – If a conditional reset is required, create a separate signal for the reset signal, and isolate this in a separate module, as shown in Example 5-18. This approach results in more readable code and improves synthesis results.

Example 5-18   Isolating conditional reset logic

Poor coding style:

```
EX18A_PROC: process ( clk, rst, a. b )
  begin
    if (rst or (a and b) = '1') then
      reg_sigs <= '0';
    elsif (clk'event and clk = '1') then
      . . .
    end if;
  end process EX18A_PROC;
```

Good coding style:

```
-- in a separate reset module
    . . .

z_rst <= rst or (a and b);
    . . .
-- in the main module
EX18B_PROC: process ( clk, z_rst)
```

```
    begin
      if (z_rst = '1') then
        reg_sigs <= '0';
      elsif (clk'event and clk = '1') then
        .  .  .
      end if;
    end process EX18B_PROC;
```

5.5   Coding for Synthesis

The following guidelines address synthesis issues. By following these guidelines, you will create code that achieves the best compile times and synthesis results, including:

- Testability
- Performance
- Simplification of static timing analysis
- Gate-level circuit behavior that matches that of the original RTL code

5.5.1   Infer Registers

Guideline – Registers (flip-flops) are the preferred mechanism for sequential logic. To maintain consistency and to ensure correct synthesis, use the following templates to infer technology-independent registers (Example 5-19 for VHDL, Example 5-20 for Verilog). Use the design's reset signal to initialize registered signals, as shown in these examples. In VHDL, do not initialize the signal in the declaration; in Verilog, do not use an `initial` statement to initialize the signal. These mechanisms can cause mismatches between pre-synthesis and post-synthesis simulation.

Example 5-19   VHDL template for sequential processes

```
    -- process with synchronous reset
    EX19A_PROC: process(clk)
      begin
        IF (clk'event and clk = '1') then
          if rst = '1' then
            .  .  .
          else
            .  .  .
          end if;
        end if;
      end process EX19A_PROC;
```

```
-- process with asynchronous reset
EX19B_PROC: process(clk, rst_a)
  begin
    IF rst_a = '1' then
      . . .
      elseif (clk'event and clk = '1') then
      . . .
    end if;
  end process EX19B_PROC;
```

Example 5-20   Verilog template for sequential processes

```
// process with synchronous reset
always @(posedge clk)
  begin : EX20A_PROC
    if (reset == 1'b1)
      begin
        . . .
      end
    else
      begin
        . . .
      end
  end // EX20A_PROC

// process with asynchronous reset
always @(posedge clk or posedge rst_a)
  begin : EX20B_PROC
    if (reset == 1'b1)
      begin
        . . .
      end
    else
      begin
        . . .
      end
  end // Ex20b_proc
```

5.5.2 Avoid Latches

Rule – Avoid using any latches in your design.

As an exception, you can instantiate technology-independent GTECH D latches. However, all latches must be instantiated and you must provide documentation that lists each latch and describes any special timing requirements that result from the latch.

Large registers, memories, FIFOs, and other storage elements are examples of situations in which D latches are permitted. Also, for 2-phase clocked synchronous RAM, you may want to use D latches to latch the address.

Note – To check your design for latches, compile the design (with no constraints for a quick compile) and use the `report_cells` command to check for latches.

Example 5-21 illustrates a VHDL code fragment that infers a latch because there is no `else` clause for the `if` statement. Example 5-22 illustrates another VHDL code fragment that infers a latch because the `z` output is not assigned for the `when others` condition.

Example 5-21   Poor coding style: Latch inferred because of missing `else` condition

```
EX21_PROC: process (a, b)
begin
  if (a = '1') then
    q <= b;
  end if;
end process EX21_PROC;
```

Example 5-22   Poor coding style: Latch inferred because of missing z output assignment

```
EX22_PROC: process (c)
begin
  case c is
    when '0' => q <= '1'; z <= '0';
    when others => q <= '0';
  end case;
end process EX22_PROC;
```

Example 5-23 illustrates a Verilog code fragment that infers latches because of missing s output assignments for the 2'b00 and 2'b01 conditions and a missing 2'b11 condition.

Example 5-23   Poor coding style: Latches inferred because of missing assignments
and missing condition

```
always @ (d)
begin
  case (d)
    2'b00: z <= 1'b1;
    2'b01: z <= 1'b0;
    2'b10: z <= 1'b1; s <= 1'b1;
  endcase
end
```

Guideline – You can avoid inferred latches by using any of the following coding
techniques:

- Assign default values at the beginning of a process, as illustrated for VHDL in
  Example 5-24.
- Assign outputs for all input conditions, as illustrated in Example 5-25.
- Use else (instead of elsif) for the final priority branch, as illustrated in
  Example 5-26.

Example 5-24   Avoiding a latch by assigning default values

```
COMBINATIONAL_PROC : process (state, bus_request)
begin
  -- intitialize outputs to avoid latches
  bus_hold <= '0';
  bus_interrupt <= '0'
  case (state) ...
  . . . . . . . . . . . . . . . .
  . . . . . . . . . . . . . . . .
end process COMBINATIONAL_PROC;
```

Example 5-25   Avoiding a latch by fully assigning outputs for all input conditions

Poor coding style:

```
EX25A_PROC: process (g, a, b)
begin
  if (g = '1') then
    q <= 0;
  elsif (a = '1') then
    q <= b;
  end if;
end process EX25A_PROC;
```

Recommended coding style:

```
EX25B_PROC: process (g1, g2, a, b)
begin
  q <= '0';
  if (g1 = '1') then
    q <= a;
  elsif (g2 = '1') then
    q <= b;
  end if;
end process EX25B_PROC;
```

Example 5-26   Avoiding a latch by using else for the final priority branch
(VHDL)

Poor coding style:

```
MUX3_PROC: process (decode, A, B)
begin
  if (decode = '0') then
    C <= A;
  elsif (decode = '1') then
    C <= B;
  end if;
end process MUX3_PROC;
```

Recommended coding style:

```
MUX3_PROC: process (decode, A, B)
begin
  if (decode = '1') then
    C <= A;
  else
    C <= B;
  end if;
end process MUX3_PROC;
```

5.5.3   Avoid Combinational Feedback

Guideline – Avoid combinational feedback; that is, the looping of combinational processes. See Figure 5-7.

BAD: Combinational processes are looped

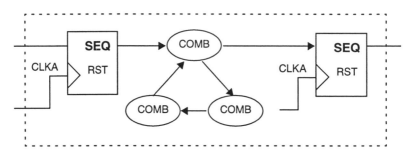

GOOD: Combinational processes are not looped

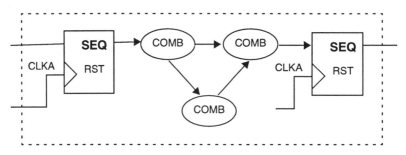

Figure 5-7   Avoiding combinational feedback

5.5.4   Specify Complete Sensitivity Lists

Rule – Include a complete sensitivity list in each of your `process` (VHDL) or
`always` (Verilog) blocks.

If you do not use a complete sensitivity list, the behavior of the pre-synthesis design
may differ from that of the post-synthesis netlist, as illustrated in Figure 5-8.

Design Compiler, as well as InterHDL's Verilint and VHDLlint, detect incomplete
sensitivity lists and issue a warning when you elaborate the design.

VHDL   **Verilog**

```
process (a)                    always @ (a)
begin                            c <= a or b;
  c <= a or b;
end process
```

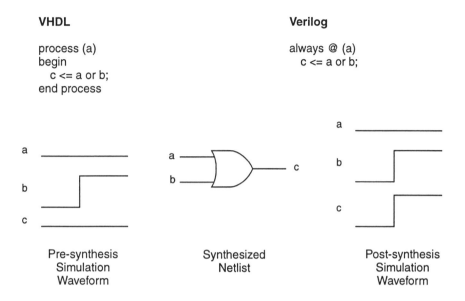

Pre-synthesis   Synthesized   Post-synthesis
Simulation   Netlist   Simulation
Waveform   Waveform

Figure 5-8   Bad example: Simulation mismatch because of incomplete sensitivity list

Combinational Blocks

For combinational blocks (blocks that contain no registers or latches), the sensitivity list must include every signal that is read by the process. In general, this means every signal that appears on the right side of an assign (<=) statement or in a conditional expression. See Example 5-27.

Example 5-27   Good coding style: Sensitivity list for combinational process block

VHDL:

```
COMBINATIONAL_PROC : process (a, inc_dec)
begin
  if inc_dec = '0' then
    sum <= a + 1;
  else
    sum <= a - 1;
  end if;
end process COMBINATIONAL_PROC;
```

Verilog:

```
always @(a or inc_dec)
begin : COMBINATIONAL_PROC
  if (inc_dec == 0)
    sum = a + 1;
  else
    sum = a - 1;
end  // COMBINATIONAL_PROC
```

Sequential Blocks

For sequential blocks, the sensitivity list must include the clock signal that is read by the process. In general, this means every control signal that appears in an `if` or `elsif` statement. See Example 5-28.

Example 5-28   Good coding style: Sensitivity list in a sequential process block

VHDL:

```
SEQUENTIAL_PROC : process (clk)
begin
  if (clk'event and clk = '1') then
    q <= d;
  end if;
end process SEQUENTIAL_PROC;
```

Verilog;

```
always @(posedge clk)
begin : SEQUENTIAL_PROC
  q <= d;
end  // SEQUENTIAL_PROC
```

Sensitivity List and Simulation Performance

Guideline – Make sure your process sensitivity lists contain only necessary signals, as defined in the sections above. Adding unnecessary signals to the sensitivity list slows down simulation.

5.5.5   Blocking and Nonblocking Assignments (Verilog)

In Verilog, there are two types of assignment statements: blocking and nonblocking. Blocking assignments execute in sequential order, nonblocking assignments execute concurrently.

Rule (Verilog only) – When writing synthesizable code, always use nonblocking assignments in `always @ (posedge clk)` blocks. Otherwise, the simulation behavior of the RTL and gate-level designs may differ.

Example 5-29 shows a Verilog code fragment that uses a blocking assignment where *b* is assigned the value of *a*, then *a* is assigned the value of *b*. The result is the circuit shown in Figure 5-9, where Register A just loops around and reassigns itself every clock tick. Register B is the same result one time unit later.

Example 5-29   Poor coding style: Verilog blocking assignment

```
always @ (posedge clk)
begin
  b = a;
  a = b;
end
```

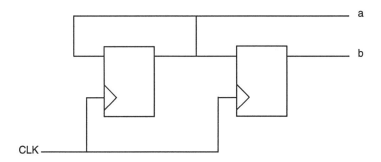

Figure 5-9   Bad example: Circuit built from blocking assignment

Example 5-30 shows a Verilog code fragment that uses a nonblocking assignment. *b* is assigned the value of *a* and *a* is assigned the value of *b* at every clock tick. The result is the circuit shown in Figure 5-10.

Example 5-30   Recommended coding style: Verilog nonblocking assignment

```
always @ (posedge clk)
begin
  b <= a;
  a <= b;
end
```

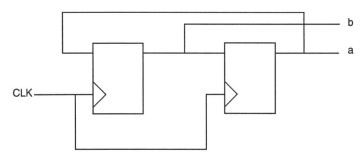

Figure 5-10   Circuit built from nonblocking assignment

5.5.6   Signal vs. Variable Assignments (VHDL)

In VHDL simulation, signal assignments are scheduled for execution in the next simulation cycle. Variable assignments take effect immediately, and they take place in the
order in which they appear in the code. Thus, they present some of the same problems
as blocking assignments in Verilog. VHDL variables are not as problematic as Verilog
blocking assignments because the interfaces between modules in VHDL are required
to be signals, so these interfaces are well-behaved. The order dependencies of variables are thus strictly local, and so it is reasonable easy to develop correct code.

Guideline (VHDL only) – When writing synthesizable code, we suggest you use
signals instead of variables to ensure that the simulation behavior of the pre-synthesis
design matches that of the post-synthesis netlist. If you feel that simulation speed will
be significantly improved by using variables, then it is certainly appropriate to do so.
Just exercise caution in creating order-dependent behavior in the code.

Example 5-31   VHDL variable assignment in synthesizable code

Poor coding style:

```
EX31_PROC: process (a,b)
variable c : std_logic;
begin
  c := a and b;
end process EX31_PROC;
```

Recommended coding style:

```
signal c : std_logic;
EX31_PROC:process (a,b)
begin
  c <= a and b;
end process EX31_PROC;
```

5.5.7   Case Statements versus if-then-else Statements

In VHDL and Verilog, a `case` statement infers a single-level multiplexer, while an `if-then-else` statement infers a priority-encoded, cascaded combination of multiplexers.

Figure 5-11 shows the circuit built from the VHDL `if-then-else` statement in Example 5-32.

Figure 5-12 shows the circuit built from the VHDL `case` statement in Example 5-33.

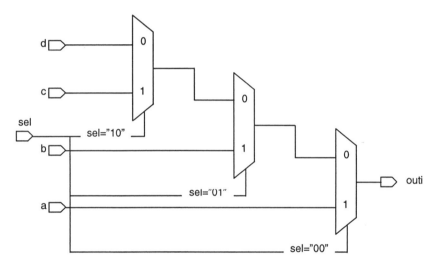

Figure 5-11   Circuit built from `if-then-else` statement

Example 5-32   Using a VHDL `if-then-else` statement

```
EX32_PROC: process (sel,a,b,c,d)
begin
  if (sel = "00") then
    outi <= a;
  elsif (sel = "01") then
    outi <= b;
  elsif (sel = "10") then
    outi <= c;
  else
    outi <= d;
  end if;
end process EX32_PROC;
```

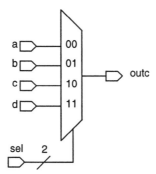

Figure 5-12   Circuit built from the case statement

Example 5-33   Using a VHDL case statement

```
EX33_PROC:process (sel,a,b,c,d)
begin
  case sel is
    when "00" => outc <= a;
    when "01" => outc <= b;
    when "10" => outc <= c;
    when others => outc <= d;
  end case;
end process EX33_PROC;
```

Guideline – The multiplexer is a faster circuit. Therefore, if the priority-encoding structure is not required, we recommend using the case statement rather than an if-then-else statement. In a cycle-based simulator, the case statement also simulates faster than the if-then-else statement.

A conditional signal assignment may also be used to infer a multiplexer. For large multiplexers, a case statement will simulate faster than a conditional assignment on most simulators, and especially cycle-based simulators. For small muxes, the relative speed of the two constructs varies with different simulators.

Example 5-34 illustrates how to use a conditional assignment to infer a mux.

Example 5-34   Using a conditional assignment to infer a mux

VHDL:

```
z1 <= a when sel_a = '1' else
      b when sel_b = '1' else
      c;

z2 <= d when sel_a = '1' else
      e when sel_b = '1' else
      f;
```

Verilog:

```
assign z1 = (sel_a) ? a : (sel_b) ? b : c;

assign z2 = (sel_a) ? d : (sel_b) ? e : f;
```

5.5.8   Coding State Machines

Observe the following guidelines when coding state machines:

Guideline – Separate the state machine HDL description into two processes, one for the combinational logic and one for the sequential logic.

Guideline – In VHDL, create an enumerated type for the state vector. In Verilog, use `'define` statements to define the state vector.

Guideline – Keep FSM logic and non-FSM logic in separate modules. See "Partitioning for Synthesis" later in this chapter for details.

For more information about coding state machines, read the Optimizing Finite State Machines chapter of the *Design Compiler Reference Manual*.

Example 5-35   VHDL FSM coding example

```
library IEEE, STD;
use IEEE.std_logic_1164.all;
use IEEE.std_logic_components.all;
use IEEE.std_logic_misc.all;
entity fsm is
  port (
    x     : in  std_logic;
    rst   : in  std_logic;
    clock : in  std_logic;
```

```
    z       : out std_logic);
end fsm;

architecture rtl of fsm is
type state is (STATE_0, STATE_1, STATE_2, STATE_3);
signal current_state, next_state : state;
begin

-- combinational process calculates next state

  COMBO_PROC : process(x, current_state)
  begin
    case (current_state) is
    when STATE_0 =>
      z <= '0';
      if x = '0' then
        next_state <= STATE_0;
      else
        next_state <= STATE_1;
      end if;
    when STATE_1 =>
      z <= '0';
      if x = '0' then
        next_state <= STATE_1;
      else
        next_state <= STATE_2;
      end if;
    when STATE_2 =>
      z <= '0';
      if x = '0' then
        next_state <= STATE_2;
      else
        next_state <= STATE_3;
      end if;
    when STATE_3 =>
      if x = '0' then
        z <= '0';
        next_state <= STATE_3;
      else
        z <= '1';
        next_state <= STATE_0;
```

```
      end if;
   when others =>
      next_state <= STATE_0;
    end case;
  end process COMBO_PROC;

-- synchronous process updates current state

  SYNCH_PROC : process(rst,clock)
  begin
    if (rst ='1') then
      current_state <= STATE_0;
    elsif (clock'event and clock ='1') then
      current_state <= next_state;
    end if;
  end process SYNCH_PROC;
end rtl;
```

Example 5-36   Verilog FSM coding example

```
module fsm(clock, rst, x, z);
input clock, rst, x;
output z;
reg [1:0] current_state;
reg [1:0] next_state;
reg z;
parameter [1:0]
  STATE_0 = 0,
  STATE_1 = 1,
  STATE_2 = 2,
  STATE_3 = 3;

// combinational process calculates next state

always @ (current_state or x)
case(current_state) //synopsys parallel_case full_case
  STATE_0 : begin
    if (x) begin
      next_state = STATE_1;
      z = 1'b0;
    end else begin
      next_state = STATE_0;
```

```
        z = 1'b0;
      end
  end
  STATE_1 : begin
    if (x)
      begin
      next_state = STATE_2;
      z = 1'b0;
      end
    else
      begin
      next_state = STATE_1;
      z = 1'b0;
      end
  end
  STATE_2 : begin
    if (x)
      begin
      next_state = STATE_3;
      z = 1'b0;
      end
    else
      begin
      next_state = STATE_2;
      z = 1'b0;
      end
  end
  STATE_3 : begin
    if (x)
      begin
      next_state = STATE_0;
      z = 1'b1;
      end
    else
        begin
        next_state = STATE_3;
        z = 1'b0;
        end
    end
    default  :  begin
        next_state = STATE_0;
```

```
        z = 1'b0;
        end
  endcase
always @ ( posedge clock or negedge rst_na)
  begin
  if (!rst_na)
    current_state = STATE_0;
  else
    current_state = next_state;
  end
endmodule
```

5.6   Partitioning for Synthesis

Good synthesis partitioning in your design provides several advantages including:

- Better synthesis results
- Faster compile runtimes
- Ability to use simpler synthesis strategies to meet timing

The following sections illustrate several recommended synthesis partitioning techniques.

5.6.1   Register All Outputs

Guideline – For each block of a hierarchical design, register all output signals from the block.

Registering the output signals from each block simplifies the synthesis process because it makes output drive strengths and input delays predictable. All the inputs of each block arrive with the same relative delay. Output drive strength is equal to the drive strength of the average flip-flop.

Figure 5-13 shows a hierarchical design in which all output signals from each block are registered; that is, there is no combinational logic between the registers and the output ports.

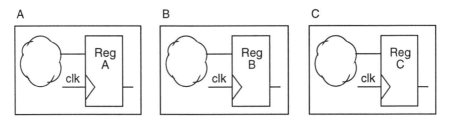

Figure 5-13   Good example: All output signals are registered

5.6.2   Locate Related Combinational Logic in a Single Module

Guideline – Keep related combinational logic together in the same module.

Design Compiler has more flexibility in optimizing a design when related combinational logic is located in the same module. This is because Design Compiler cannot move logic across hierarchical boundaries during default `compile` operations.

Figure 5-14 shows an example design where the path from register A to register C is split across three modules. Such a design inhibits Design Compiler from efficiently optimizing the combinational logic because it must preserve the hierarchical boundaries in the design.

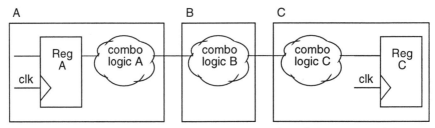

Figure 5-14   Bad example: Combinational logic split between modules

Figure 5-15 shows a similar design in which the related combinational logic is grouped into a single hierarchical block. This design allows Design Compiler to perform combinational logic optimization on the path from register A to register C.

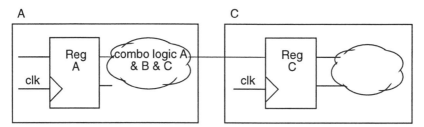

Figure 5-15   Better example: Combinational logic grouped into same module

Figure 5-16 shows an even better design where the combinational logic is grouped into the same module as the destination register. This design provides for improved sequential mapping during optimization because no hierarchical boundaries exist between the sequential logic and the combinational logic that drives it.

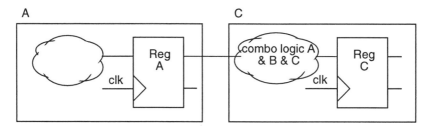

Figure 5-16   Best example: Combinational logic grouped with destination register

Keeping related combinational logic in the same module also eases time budgeting and allows for faster simulation.

5.6.3   Separate Modules That Have Different Design Goals

Guideline – Keep critical path logic in a separate module from noncritical path logic so that Design Compiler can optimize the critical path logic for speed, while optimizing the noncritical path logic for area.

Figure 5-17 shows a design where critical path logic and noncritical path logic reside in the same module. Optimization is limited because Design Compiler cannot perform different optimization techniques on the two groups of logic.

ModuleA

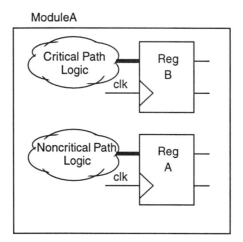

Figure 5-17   Bad example: Critical path logic grouped with noncritical path logic

Figure 5-18 shows a similar design where the critical path logic is grouped into a separate module from the noncritical path logic. In this design, Design Compiler can perform speed optimization on the critical path logic, while performing area optimization on the noncritical path logic.

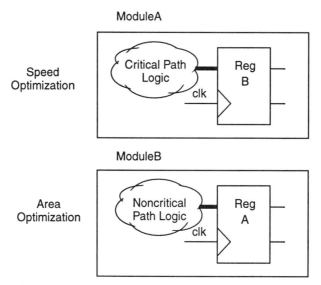

Figure 5-18   Good example: Critical path logic and noncritical path logic
grouped separately

5.6.4   Asynchronous Logic

Guideline – Avoid asynchronous logic.

Asynchronous logic is more difficult to design correctly and to verify. Correct timing and functionality may be technology dependent, which limits the portability of the design.

Guideline – If asynchronous logic is required in the design, partition the asynchronous logic in a separate module from the synchronous logic.

Isolating the asynchronous logic in a separate module makes code inspection much easier. Asynchronous logic need to be reviewed carefully to verify its functionality and timing.

5.6.5   Arithmetic Operators: Merging Resources

A resource is an operator that can be inferred directly from an HDL, as shown in the following code fragment:

```
if ctl = '1' then
  z <= a + b;
else
  z <= c + d;
end if;
```

Normally, two adders are created in this example. If only an area constraint exists, however, Design Compiler is likely to synthesize a single adder and to share it between the two additions. If performance is a consideration, the adders may or may not be merged.

For Design Compiler to consider resource sharing, all relevant resources need to be in the same level of hierarchy; that is, within the same module.

Figure 5-19 is an example of poor partitioning. In this example, resources that can be shared are separated by hierarchical boundaries.

Figure 5-20 is an example of good partitioning because the two adders are in the hierarchy level. This partitioning allows Design Compiler full flexibility when choosing whether or not to share the adders.

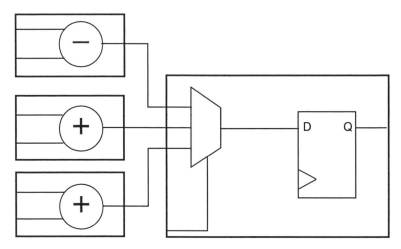

Figure 5-19   Poor partitioning: Resources area separated by hierarchical boundaries.

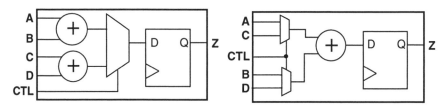

Figure 5-20   Good partitioning: Adders are in the same hierarchy

5.6.6   Partitioning for Synthesis Runtime

In the past, most synthesis guidelines have recommended keeping modules relatively small in order to reduce synthesis runtime. Improvements to Design Compiler, increases in workstation performance, and more experience with large designs has changed this.

The most important considerations in partitioning should be the logic function, design goals, and timing and area requirements. Grouping related functions together is much better than splitting functions artificially, and creating complex inter-block timing dependencies. Good timing budgets and appropriate constraints can have a larger impact on synthesis runtime than circuit size. In one test case, synthesis went from nine hours to 72 hours when the critical range was increased from 0.1 ns to 10 ns.

By grouping logic by design goals, the synthesis strategy can be focused, reducing synthesis runtime. For example, if the goal for a particular block is to minimize area, and timing is not critical, then the synthesis scripts can be focused on area only, greatly reducing runtime.

Overconstraining a design is one of the biggest causes of excessive runtime. A key technique for reducing runtimes is to develop accurate timing budgets early in the design phase and design the macro to meet these budgets. Then, develop the appropriate constraints to synthesize to this budget. Finally, by developing a good understanding of the Design Compiler commands that implement these constraints, you can achieve an optimal combination of high quality of results and low runtime.

For more information on synthesizing large designs, including the test case mentioned above, see "Synthesis Methodology for Large Designs - Design Compiler 1997.01 Release" from Synopsys.

5.6.7   Avoid Point-to-Point Exceptions and False Paths

A point-to-point exception is a path from the output of one register to the input of another that does not follow the standard objective of having the data traverse the path in one clock cycle. A multicycle path is the prime example of a point-to-point exception.

Multicycle paths are problematic because they are more difficult to analyze correctly and lend themselves to human error. They must be marked as exceptions to the static timing analysis tool; it is all too easy to mark a path as an exception by mistake and not perform timing analysis. Most static timing analyzers work much better on standard paths than on exceptions.

Guideline – Avoid multicycle paths in your design.

Guideline – If you must use a multicycle path in your design, keep point-to-point exceptions within a single module, and comment them well in your RTL code.

Isolating point-to-point exceptions (for example, multicycle paths) within a module improves compile runtime and synthesis results. Also, the `characterize` command has limited support for point-to-point exceptions that cross hierarchical boundaries.

Figure 5-21 shows a good partitioning example where the start and end points of a multicycle path occur within a single module.

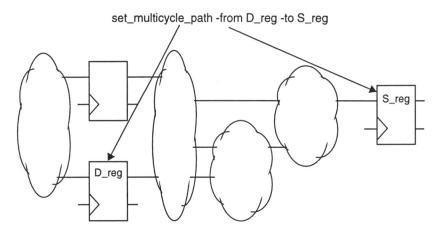

Figure 5-21   Good example: Isolating a point-to-point exception to a single module

Guideline – Avoid false paths in your design.

False paths are paths that static timing analysis identifies as failing timing, but that the designer knows are not actually failing.

False paths are a problem because they require the designer to ignore a warning message from the timing analysis tool. If there are many false paths in a design, it is easy for the designer to accidently ignore valid warning message about actual failing paths.

5.6.8   Eliminate Glue Logic at the Top Level

Guideline – Do not instantiate gate-level logic at the top level of the design hierarchy.

A design hierarchy should contain gates only at leaf levels of the hierarchy tree. For example, Figure 5-22 shows a design where a NAND gate exists at the top level, between two lower-level design blocks. Optimization is limited because Design Compiler cannot merge the NAND with the combinational logic inside block C.

Top

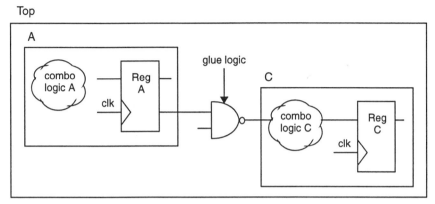

Figure 5-22   Bad example: Glue logic existing at top level

Figure 5-23 shows a similar design where the NAND gate is included as part of the combinational logic in block C. This approach eliminates the extra CPU cycles needed to compile small amount of glue logic and provides for simpler synthesis script development. An automated script mechanism only needs to compile and characterize the leaf-level cells.

Top

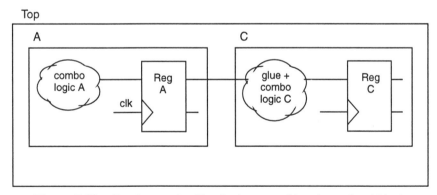

Figure 5-23   Good example: Glue logic grouped into lower-level block

5.6.9   Chip-Level Partitioning

Figure 5-24 shows the partitioning recommendation for the top-level of an ASIC.

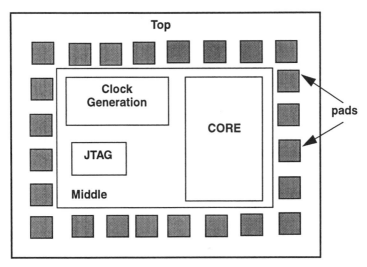

Figure 5-24   Top-level partitioning for an ASIC

Guideline – Make sure that only the top level of the design contains an I/O pad ring. Within the top level of hierarchy, a middle level of hierarchy contains IEEE 1149.1 boundary scan (JTAG) modules, clock generation circuitry, and the core logic. The clock generation circuitry is isolated from the rest of the design as it is normally hand crafted and carefully simulated. This hierarchy arrangement is not a requirement, but allows easy integration and management of the test logic, the pads, and the functional core.

5.7   Designing with Memories

Memories present special problems for reusable design, since memory design tends to be foundry specific. Macros must be designed to deal with a variety of memory interfaces. This section outlines some guidelines for dealing with these issues, in particular, designing with synchronous and asynchronous memories.

Synchronous memories present the ideal case, and their interfaces are in the general form shown in Figure 5-25. Figure 5-26 shows the equivalent asynchronous RAM design.

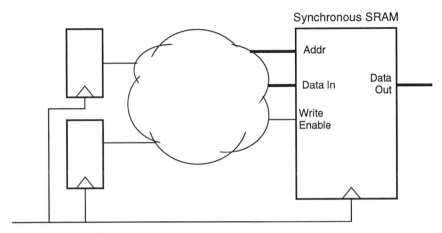

Figure 5-25   Synchronous memory interface

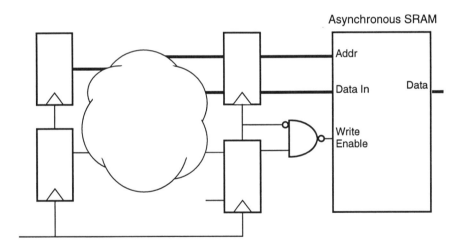

Figure 5-26   Asynchronous memory interface

Guideline – Partition the address and data registers and the write enable logic in a separate module. This allows the memory control logic to work with both asynchronous and synchronous memories. See Figure 5-27.

In the design shown in Figure 5-27, the interface module is required only for asynchronous memories. The functionality in the interface module is integrated into the synchronous RAM.

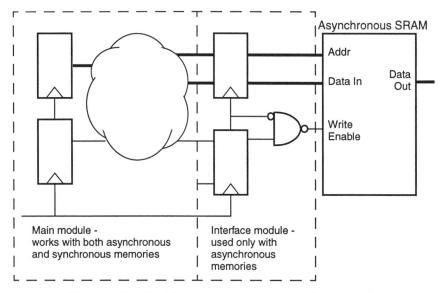

Figure 5-27   Partitioning memory control logic separately

5.8   Code Profiling

In some cases, *code profiling* can assist you in optimizing your code. Some simulators, and several third-party code coverage tools, provide the capability of tracking how often each line of code is executed during a given simulation run.

Profiling is a valuable tool that can reveal bottleneck areas in the model. However, you must keep in mind that the profiler looks only at the frequency with which a line is executed, not at how expensive that construct is in terms of machine cycles. For example, performing a variable assignment statement differs a great deal from performing a signal assignment.

Code coverage tools that measure path coverage as well as statement coverage can be very useful for analyzing how well a given test vector set exercises the model and for checking redundancies in the model itself. For example, if some parts of the model receive no execution coverage at all, either the vectors are failing to exercise the model fully or that portion of the model is redundant. See Chapter 7 for more discussion of code coverage tools.

CHAPTER 6

Macro Synthesis Guidelines

This chapter discusses strategies for developing macro synthesis scripts that enable the integrator to synthesize the macro and meet timing goals. The topics include:

- Overview of the synthesis problem
- Synthesis strategies for reusable macros
- High-performance synthesis
- RAM and datapath generators
- Coding guidelines for synthesis scripts

6.1 Overview of the Synthesis Problem

There are some special problems associated with the synthesis of parameterizable soft macros:

- The macro and synthesis scripts must allow the integrator to synthesize the macro and meet timing goals in the final chip.
- The macro must meet timing with the integrator's gate array or standard cell library.
- The macro must meet timing in the integrator's specific configuration of the macro.

This chapter presents a set of tools and methodologies for achieving these goals.

The synthesis guidelines in this chapter are based on many of the same fundamental principles guiding the previous chapter. First and foremost, synthesis and timing design must start at the beginning of the macro design cycle.

That is:

- Functional specifications for the macro must describe the timing, area, wireload model, and power requirements for the design.
- Detailed technical specifications for the macro and its various subblocks must describe the timing requirements and interfaces in detail, including specifications for input and output delays.
- RTL needs to be coded from the outset to meet both the functional and the timing requirements of the design. Coding for functionality first, and then fixing timing problems later, causes significant delays and poor overall performance in many designs.

If these fundamental guidelines are followed, then synthesis is a straightforward issue. Each synthesizable unit or module in the design has a timing budget. Once each module meets this timing budget, the macro is assured of meeting its overall timing goals. Synthesis problems become localized, so the difficult problems can be solved on small modules, where they are the most tractable.

6.2   Macro Synthesis Strategy

The recommended synthesis strategy for macros is to develop a set of constraints for the macro early in the design process and to use a bottom-up synthesis strategy.

6.2.1   Macro Timing Budget

Rule – The basic timing budget for the macro must be developed as part of the specification process, before the design is partitioned into blocks and before coding begins. This timing budget must be reviewed regularly during the design process to ensure that it is still reasonable and consistent.

The macro timing budget must specify:

- Clock definition
- Setup time requirements for all signals going into the macro
- Clock to output delay requirements for all synchronous outputs of the macro
- Input and output delays for all combinational paths through the macro
- Loading budget for outputs and driving cell for inputs

- Operating conditions, including temperature and voltage

Note that combinational paths through the macro are discouraged, because they create non-local synthesis problems that can be very difficult to resolve. The combinational paths must be carefully documented and their timing budgets closely examined to make sure the design constraints can be met. The preferred method for specifying these combinational delays is to specify the input arrival times and the required output time with respect to the clock, assuming the clock is present in the block

6.2.2 Subblock Timing Budget

Rule – The basic timing budget must be developed for each subblock in the macro. This budget must be developed at the time that the design is partitioned into subblocks, and before coding begins. The budget must be reviewed regularly during the design process to ensure that it is still reasonable and consistent.

The subblock timing budget must specify:

- Clock definition
- Wireload model
- Setup time requirements for all signals going into the subblock
- Clock to output delay requirements for all synchronous outputs of the subblock
- Input and output delays for all combinational paths through the subblock
- Loading budget for outputs and driving cell for inputs
- Operating conditions, including temperature and voltage

A good nominal starting point for the loading and driving specifications is to use a two-input NAND gate as the driving cell and a flip-flop data input pin as the output load.

As above, combinational paths through subblocks are discouraged. In our experience, most synthesis problems arise from these combinational paths.

6.2.3 Synthesis in the Design Process

Synthesis starts as the individual designers are developing the subblocks of the macro, and is initially performed with a single technology library. Later, during the productization phase, the entire macro is synthesized to multiple libraries to ensure portability.

The designer should start running synthesis as soon as the RTL passes the most basic simulation tests. Performing synthesis at this stage allows the designer to identify and fix timing problems early. Because fixing the tough timing problems usually means modifying or restructuring the RTL, it is much better to deal with these problems before the code is completely debugged.

Early synthesis also allows the designer to identify the incremental timing costs of new functionality as it is added to the code.

The target at this early stage of synthesis should be to get within about 10-20% of the final timing budget. This should be close enough to assure that the RTL code is structured correctly. The additional effort to achieve the timing budget completely is not worth the effort until the code is passing all functional tests. This additional effort will most likely consist of modifying the synthesis scripts and refining the timing budgets.

The subblocks should meet all timing budgets, as well as meeting all functional verification requirements, before being integrated into the macro.

6.2.4   Subblock Synthesis Process

Guideline – The subblock synthesis process consists of three phases:

1. Perform a compile on the subblock, using constraints based on the budget.
2. Perform a characterize-compile on the whole subblock, to refine the timing constraints and re-synthesize the subblock.
3. Iterate if required.

The characterize-compile strategy in step 2 is documented in Appendix C of the Fundamentals section of the *Design Compiler Reference Manual*.

6.2.5   Macro Synthesis Process

When the subblocks are ready for integration, we are ready to perform macro-level synthesis.

Guideline – The macro synthesis process consists of three phases:

1. Perform a compile on each of the subblocks, using constraints based on the budget.
2. Perform a characterize-compile on the whole macro to improve timing and area.
3. If necessary to meet the timing goals, perform an incremental compile.

The characterize-compile in step 2 is needed to develop accurate estimates of the loading effects on the inputs and outputs of each subblock. Initially, the drive strength of the cells driving inputs, and the loading effects of cells driven by the outputs, are estimated and set manually. The `set_driving_cell` and `set_load` commands are used for this purpose. The characterize-compile step derives actual drive strengths and loading from the rest of the macro. Clearly, this requires an initial synthesis of the entire macro in order to know what cells are driving/loading any specific subblock input/output.

6.2.6   Wireload Models

Wireload models estimate the loading effect of metal interconnect upon cell delays. For deep submicron designs, this effect dominates delays, so using accurate wireload models is critical.

The details of how a given technology library does wireload prediction varies from library to library, but the basic principles are the same. A statistical wire length is determined based on the physical size of the block. From this statistical wire length and the total input capacitance of the nodes on the net, the synthesis tool can determine the total loading on the driving cell.

The most critical factor in getting an accurate statistical wire length is to estimate accurately the size of the block that will be placed and routed as a unit. Typically, a macro will be placed and routed as a single unit, and the individual subblocks that make up the macro will be flattened within the macro. Thus, the appropriate wireload model is determined by the gate count (and thus area) of the entire macro at the top level.

When we synthesize a subblock, we must use the wireload model for the full macro, not just the subblock. If we use just the gate count of the subblock to determine the wireload model, we will get an optimistic model that underestimates wire delays. When we then integrate the subblocks into the macro and use the correct wireload model, we can run into significant timing problems.

6.2.7   Preserve Clock and Reset Networks

Clock networks are typically not synthesized; we rely on the place and route tools to insert a balanced clock tree with very low skew. Asynchronous reset networks are also typically treated as special networks, with the place and route tools inserting the appropriate buffers. These non-synthesized networks need to be identified to the synthesis tool.

Guideline – Set `dont_touch_network` on clock and asynchronous reset networks. Include the required `dont_touch_network` commands in the synthesis scripts for the design. See Example 6-1.

Example 6-1 Using `dont_touch_network` on clocks and reset networks

```
dont_touch_network {clk rst}
```

6.2.8   Code Checking Before Synthesis

Several checks should be run before synthesis. These checks can spot potential synthesis problems without having to perform a complete compile.

Lint Tools

Lint-like tools (InterHDL's Verilint and VHDLlint, for example) can quickly check for many different potential problems, including:

- Presence of latches
- Non-synthesizable constructs like "`===`" or `initial`
- Whether a `case` statement was inferred as a mux or a priority encoder
- Whether all bits in a bus were assigned
- Unused macros, parameters, or variables

dc_shell

Once the RTL passes lint, the elaboration reports from Design Compiler should be examined to check:

- Whether sequential statements were inferred as flip-flops or latches
- Whether synchronous or asynchronous reset was inferred

A clean elaboration of the design is a critical first step in performing synthesis.

6.2.9   Code Checking After Synthesis

After synthesis, a number of Design Compiler checks can be run to identify potential problems:

Loop Checking
Run `report_timing -loops` to determine whether there are any combinational loops.

Checking for Latches
Run `all_registers -level_sensitive` to get a report on latches in the design.

Check for Design Rule Violations

Run `check_design` to check for missing cells, unconnected ports, and inputs tied high or low.

Verify Testability

Run `check_test` to verify that there are scan versions of all flops, and to check for any untestable structures. Soft macros are typically not shipped with scan flops inserted because scan is usually done on a chip-wide basis rather than block-by-block. Thus, it is essential to verify that scan insertion and automatic test pattern generation (ATPG) will be successful.

As part of the productization phase of the macro development process, full ATPG is run.

Verify Synthesis Results

Use Formality to verify that the RTL and the post-synthesis netlist are functionally equivalent.

6.3 High-Performance Synthesis

As chip size and complexity increase, it becomes more critical to have some interactivity between the synthesis and layout phases of chip design. Currently, some alternatives to the standard sequence are becoming available.

6.3.1 Classical Synthesis

In standard ASIC design, the synthesis phase has no automated interaction with the subsequent layout phase. Synthesis must generate the netlists without any feedback from floorplanning and place-and-route tools, and there is no opportunity to modify synthesis based on findings during layout. Hand iteration between synthesis and placement is slow and painful. If resynthesis is necessary, layout generally has to be redone from scratch. While this lack of interactivity between the synthesis and layout stages is manageable for smaller chip sizes, it is problematic and distinctly not optimal for today's large System-on-a-Chip designs.

The problems produced by this lack of interactivity between synthesis and layout are exacerbated because, as transistors and cells become faster, cell delays decrease and the percentage of delay due to loading factors increases. Information about physical placement becomes more important for synthesis.

6.3.2   High-Performance Synthesis

New tools, such as Synopsys' Links-to-Layout and Floorplan Manager, provide inter-activity between the synthesis and placement phases of design. Such tools allow *high-performance synthesis* by forward-annotating constraints such as timing and net priorities to a floorplanner or place and route tool, and back-annotating physical information such as net delays, net capacitance, and physical grouping to the synthesis tool. This interactivity greatly improves the speed and accuracy of synthesis and layout by speeding the iterations, and because synthesis and layout are both performed with actual values rather than estimates.

6.3.3   Tiling

In some cases, the designer knows that certain elements will fit together—"tile"—into a compact pattern that can then be repeated. Floorplanning and place and route tools are not likely to detect the possibility of such compact configurations. Historically, the designer has had to lay out such areas by hand, and then provide the floorplanner with a "black box" macro for these areas. Such hand crafting produces highly compact layout, but is costly in terms of time spent. Tools for automating this hand-crafted tiling process are becoming available.

6.4   RAM and Datapath Generators

Memories and datapaths present a special set of problems for design reuse. Historically, memories and high performance datapaths have been designed at the physical level, making them very technology dependent.

6.4.1   Memory Design

There is almost no logical design content to (most) memory design. There are single port vs. multi-port and synchronous vs. asynchronous memories, but what truly differentiates a good memory design from a bad one is the size, power, and speed of the memory. The extreme regularity of memory determines the design methodology. A memory cell is developed, hopefully as small and fast and low power as possible. The memory cell is then replicated in a regular tiling fashion. Unfortunately, the optimal cell design is very dependent on the underlying fabrication process. Thus, each silicon vendor has tended to develop a unique memory compiler tailored to the specific requirements of the target silicon technology.

The result is that memory designs are not portable or reusable. This situation places a significant burden on the developer of reusable designs. In Chapter 5, we described some approaches for dealing with memories in designing reusable macros, and later in this chapter we describe the integration flow for using macros with memory modules in chip-level designs. But first, we discuss datapath design, which, until recently, shared many of the same problems as memory design.

6.4.2   Datapath Design

In those datapaths that are dominated by arithmetic functions, the functionality of the design is usually straightforward. The functionality of a 32-bit multiply-accumulate block, for example, is clear and does not help differentiate a design. In order to have a 32-bit MAC that is superior to a competitor's, it is necessary to exploit hardware structure to achieve a faster, smaller, or lower-power design. Historically, this approach has led to tools and methodologies designed to exploit the structural regularity in the datapath, and thus derive a superior physical layout.

Datapath Design Issues

There are three major problems with traditional approaches to datapath design. First, irregular structures like Wallace tree multipliers can outperform regular structures. Second, the datapath designs produced are not portable to new technologies and do not lend themselves to reuse. Third, great majority of modern applications are poor candidates for the traditional approach, which is best suited to datapaths that are relatively simple (few number of operations) and highly regular (uniform bit-widths).

If we look at the history of datapath design, a typical datapath in 1988 would be a simple, regular datapath, such as a CPU ALU. Regular structures like muxes and adders dominated; bit slicing was used extensively, and was effective in deriving dense, regular layouts. A 32-bit MAC was a separate chip.

In 1998, graphics, video, and digital signal processing applications are the most common datapath designs. Blocks like IDCTs, FIRs, and FFTs are common datapath elements, and a 32-bit MAC is just a small component in the datapath. The significant increase in applications for complex datapaths, along with intense pressures to reduce development time, has resulted in a desire to move datapath design to a higher level of design abstraction as well as to leverage design reuse techniques.

Datapath design tools and methodologies are rapidly evolving to meet this need.

Datapath Design Tool Evolution

In the past, designers predominately designed datapaths by handcrafting the design. They captured structural information about the design in schematics and then developed a physical layout of the design. The physical design was laid out for a single bit-slice of the datapath, then replicated. For regular datapaths dominated by muxes and adders, this approach produced dense, regular physical designs. These handcrafted designs exhibit:

- High performance because the methodology effectively exploited the regular structure of the logic
- Low productivity because of the amount of handcrafting required
- Poor portability because the results were design and technology specific

In the more recent past, designers started using layout-oriented datapath design tools. With these tools, structural descriptions of the design are entered either in schematic form or in HDL, but with severe constraints limiting the subset of the language that can be used. These tools automate much of the handcrafting that was done before, such as developing bit-slice layouts and regular structures. The designs result in:

- High performance for regular structures
- Poor performance for irregular, tree-based structures like Wallace-tree multipliers
- Poor performance for structures with varying bit widths, a common characteristic of graphics designs such as IDCTs, digital filters, or any other design employing techniques like saturation, rounding, or normalization
- Moderate productivity because of the automation of design tasks
- Poor portability because designs were still design and technology specific

A number of datapath designers have used conventional synthesis to improve the technology portability of their designs. Conventional synthesis uses generic operators with structures that are expressed in a generic library; during synthesis, the designed is then mapped onto the specific technology library. Designs using conventional synthesis have:

- Moderate performance for complex datapaths, very good performance on simple ones
- Moderate productivity for complex datapaths, very good productivity on simple ones
- Good portability

Today's most advanced datapath design tools, like Module Compiler, approach the problem of datapath design differently. With these tools, the designer enters the structural description for the design in a flexible HDL. The tool then performs synthesis to generate an optimal netlist for the entire datapath. The designer has the flexibility to manipulate the structural input to guide or control the synthesis. Because the functionality for even relatively complex datapaths is well known, the designer can focus on the implementation structure to differentiate the datapath design solution.

The key realization behind these new tools is that good datapath design starts with a good netlist, not with a bit-slice physical design. Today's datapaths are dominated by tree structures that have little of the regularity of earlier datapaths. For these structures, automatic place and route tools do at least as good a job as hand design, and often better. The key to performance is to develop the best possible detailed structure (the netlist) and then map it onto the available technology (through place and route). And unlike conventional synthesis, these specialized tools use algorithms that are specific for datapath synthesis, producing better netlists in shorter time.

One of the key benefits of these tools is that they are significantly faster than any other design method, often an order of magnitude or more faster than conventional synthesis. One advantage of this speed is that many different implementations for the design can be developed and evaluated. For instance, when designing an IDCT, the designer can experiment with different saturation algorithms, different numbers of multipliers, and different numbers of pipelines. As a result of this exploration, a superior architecture can be developed. This improved architecture can more than compensate for any loss in performance compared to a handcrafted design.

Because they allow superior, technology-independent designs, these tools provide the first opportunity to develop reusable datapath designs without sacrificing performance. This capability is essential for the design of reusable blocks for complex chips in datapath-intensive domains such as video, graphics, and multimedia.

With these tools, designs have:

- High performance – implementation exploration allows superior designs
- High productivity – extremely fast synthesis times allow very rapid development of very complex designs
- High portability – because the source is technology independent and can be parameterized, it is very portable across technologies and from design to design

The next step in the evolution of datapath tools is to extend these specialized synthesis tools to include links to physical design. Although irregular structures tend to dominate most large datapaths today, there are still many designs that have substantial portions that are very regular. Exploiting this regularity could even further improve the performance of datapath circuits. Also, like any synthesis tool, links to physical design can help improve circuit performance and reduce the iterations between logic design and physical design.

6.4.3   Design Flow Using Module Compiler

Module Compiler (MC) is a Synopsys datapath synthesis and optimization tool that provides an alternative method of designing and synthesizing complex arithmetic datapaths. For such datapaths, MC offers better quality of results and much faster compile times than general purpose synthesis tools. The compile times are so much faster (1-2 orders of magnitude) than standard synthesis that it is possible to quickly code and synthesize alternative implementations. Designers can quickly evaluate tradeoffs between timing, power, and area, and converge on optimal designs much faster than by conventional hand-crafting or general RTL-based synthesis.

In Module Compiler, you describe the datapath in the Module Compiler Language (MCL), a Verilog-like datapath description language. MC produces:

- A Verilog or VHDL gate-level netlist
- A Verilog or VHDL simulation model
- Area and timing reports
- Placement guidance information for layout

Some of Module Compiler's highlights are:

Interfaces
Module Compiler supports both GUI and a command-line modes.

Inputs
The inputs are a high-level description of the datapath in MCL and some design constraints. MCL has the look and feel of the Verilog hardware description language, but is better suited to the task of describing the synthesis and optimization of datapaths. The design constraints can be entered from the GUI or embedded in the description.

Work flow
MC is designed to support two work flows: the "exploration loop" and the "debugging loop" (Figure 6-1). The two flows are typically interleaved, with one feeding into the other.

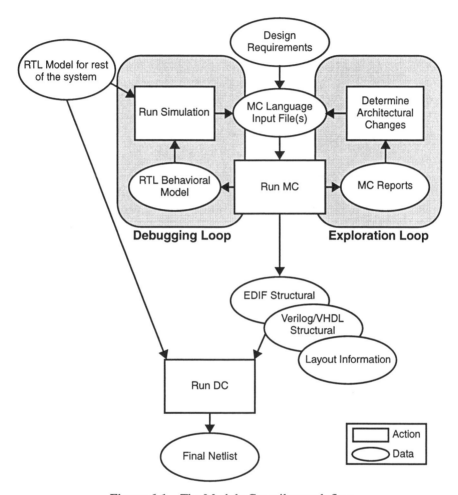

Figure 6-1   The Module Compiler work flow

The typical MC work flow is as follows:

- In the exploration loop, the designer explores the timing and area performance of alternate datapath designs. The designer codes the prospective datapath and uses MC to synthesize the input and generate reports on area, timing, and power. The designer uses these reports to optimize the macro architecture.

- In the debugging loop, MC synthesizes and optimizes the input and generates a Verilog or VHDL behavioral model and netlist. Simulation on these outputs confirms the accuracy of the network description and that latency is acceptable.

- After exploration and debug are completed, the designer uses MC to generate the final datapath netlist.

• If the datapath is part of a larger design, the designer reads both the datapath netlist and the RTL for the rest of the design in to the synthesis tool. The datapath netlist can be "dont_touch'ed" so that no additional optimizations are performed it. This option results in the fastest compile time. On the other hand, the designer can have the synthesis tool re-optimize the netlist. On some designs, some incremental improvement in timing and/or area can be achieved by this approach.

6.4.4   RAM Generator Flow

The typical RAM generator work flow, shown in Figure 6-2, is similar to that of datapath generators such as Module Compiler. With RAM compilers, the designer:

• Describes the memory configuration, through either a GUI or a command-line interface. The designer selects the family of memory, typically trading off power and area versus speed.

• Invokes the memory compiler, which produces a simulation model and a synthesis model for the memory.

• Performs simulation with models for the rest of the system to verify the functionality of the memory interfaces.

• Performs synthesis with the synthesis model for the RAM and the RTL for the rest of the system. The synthesis model for the RAM is key in determining overall chip timing and allowing optimal synthesis of the modules that interface to the RAM.

6.4.5   Design Reuse with Datapath and RAM Compilers

The input to a GUI on a RAM generator is not reusable by itself. However, the generator is a reuse tool. Most of the knowledge required to design the RAM resides in the tool, not the inputs to the tool. It is so easy to create new configurations using the RAM compiler that memory design becomes very straightforward for the chip designer. The difficult aspects of RAM design have all been encapsulated by the tool and are hidden from the user.

Module Compiler provides similar reuse capabilities. By encapsulating the difficult parts of datapath design, such as adder and multiplier tree structures and merging of arithmetic operators, MC reduces the input requirements for describing the datapath to an absolute minimum. The tool itself is the key to datapath reuse.

Unlike RAM compilers, however, the MCL code describing the datapath does have a significant design content. This code can be reused for many designs. One of the strengths of an encapsulating tool like MC is that the datapath description in MCL code is extremely simple and easy to understand. These features, of course, are the fundamental requirements for reusability.

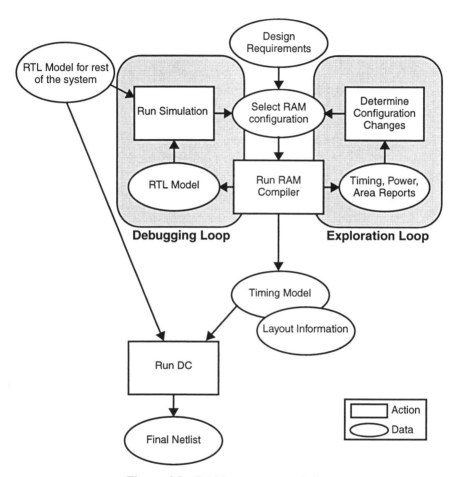

Figure 6-2 RAM generator work flow

RAM compilers and datapath compilers like MC offer a challenge to the design reuse community: Are there other domains where sufficient design expertise can be encapsulated in a tool, so that significant design reuse can be obtained from the tool itself?

6.5   Coding Guidelines for Synthesis Scripts

Many of the coding guidelines described in Chapter 5 apply equally well to all scripts, including synthesis scripts.

The following rules and guidelines apply particularly to synthesis scripts:

Rule – All scripts, including synthesis scripts, should begin with a header describing the file, its purpose, its author, and its revision history.

Rule – Comments should be used extensively to describe the synthesis strategy being executed.

Rule – All scripts used to build the design should be under revision control and a bug tracking system, just as the source code is.

Guideline – Keep the line length to 72 characters or less.

Lines that exceed 80 characters are difficult to read in print and on standard terminal width computer screens. The 72 character limit provides a margin that enhances the readability of the code.

For dc_shell commands, use a backslash (\) to continue the statement onto the next line if the command exceeds 72 characters and begin the next line with an indent.

Rule – No hard-coded numbers, data values, or filenames should be buried in the body of the script. Variables should be used in the body of the script and their values set at the top of the script.

Rule – No hard-coded paths should appear in any scripts. Scripts with hard-coded paths are not portable because hard-coded paths prevent the script from being reused in other environments.

Rule – Scripts should be as simple as they can be and still meet their objectives. Synthesis scripts that use only the most common commands are more easily understood and modified.

Rule – Common commands such as those defining the library and search paths should reside in a single setup file for the design, usually in the .synopsys_dc.setup file or in a file that can be included in .synopsys_dc.setup. All other synthesis scripts should perform only those unique tasks for which they were written. Having libraries or search paths defined in multiple files makes modification difficult.

Rule – Synthesis scripts for parameterized soft macros need to be tested as thoroughly as any source code. In particular, all statements and all paths through the script must be tested. Some scripting bugs appear only when the script is used to compile the macro in a particular configuration; these bugs must be uncovered before shipping the script to a customer.

Guideline – Run the syntax checker on Design Compiler scripts before running the script. The syntax checker can spot many of the scripting errors that can cause DC to halt or produce useless results.

The following example shows how to use the Design Compiler syntax checker:

```
dc_shell -syntax_check -f ./scripts/my_compile.scr
```

CHAPTER 7

Macro Verification Guidelines

This chapter discusses issues in simulating and verifying macros, including the importance of reusable testbenches and test suites. The topics are:

- Overview of macro verification
- Testbench design
- RTL testbench style guide
- Timing verification

7.1 Overview of Macro Verification

Design verification is consistently one of the most difficult and challenging aspects of design. Parameterized, soft macros being designed for reuse present some particular challenges:

- The verification goal must be for zero defects because the macro may be used in anything from a computer game to a mission-critical aerospace application.
- The goal of zero defects must be achieved for all legal configurations of the macro, and for all legal values of its parameters.
- The integration team must be able to reuse the macro-level testbenches and test suites because the macro must be verified both as a standalone design and in the context of the final application.
- Because the macro may be substantially redesigned in the future, the entire set of testbenches and test suites must be reusable by other design teams.

- Because some testbenches may be used in system testing, the testbenches must be compatible with the simulators and emulators used throughout the system testing process.

7.1.1   Verification Plan

Because of the inherent complexity and scope of the functional verification task, it is essential that comprehensive functional validation plans be created, reviewed, and followed by the design team. By defining the verification plan early, the design team can develop the verification environment, including testbenches and verification suites, early in the design cycle. Having a clear definition of the criteria that the macro verification must meet before shipment helps to focus the verification effort and to clarify exactly when the macro is ready to ship.

The specific benefits of developing a verification plan early in the design cycle include:

- The act of creating a functional validation plan forces designers to think through what are typically very time-consuming activities prior to performing them.
- A peer review of the functional validation plan allows a pro-active assessment of the entire scope of the task.
- The team can focus efforts first on those areas in which validation is most needed and will provide the greatest payoff.
- The team can minimize redundant effort.
- The engineers on the team can leverage the cumulative experience and knowledge of the entire team.
- A functional validation plan provides a formal mechanism for correlating project requirements to specific validation tests, which, in turn, allows the completeness (coverage) of the test suite to be assessed.
- Early identification of validation tests allows their development to be tracked and managed more effectively.
- A functional validation plan may serve as documentation of the validation tests and testbench - a critical element for the re-use of these items during regression testing and on subsequent projects. This documentation also reduces the impact of unexpected personnel changes midstream during a project.
- The information contained in the functional validation plan enables a separate validation support team to create a verification environment in parallel with the design capture tasks performed by the primary design team. This can significantly reduce the design cycle time.

The verification environment is the set of testbench components such as bus functional models, bus monitors, memory models, and the structural interconnect of such components with the design-under-test. Creation of such an environment may involve in-house development of some components and/or integration of off-the-shelf models.

The verification plan should be fully described either in the functional specification for the macro or in a separate verification document. This document will be a living document, changing as issue arise and strategies are refined. The plan should include:

- A description of the test strategy, both at the subblock and the top level.
- A detailed description of the simulation environment, including a block diagram.
- A list of testbench components. For each component, there should be a summary of key required features. There should also be an indication of whether the component already exists, can be purchased from a third party, or needs to be developed.
- A list of required verification tools, including simulators.
- A list of specific tests, along with the objective and estimated size of each. The size estimate can help develop an estimate of the effort required to develop the test.
- An analysis of the key specifications of the macro, and identification of which tests verify each of these specifications.
- A specification of what functionality of the macro will be verified at the subblock level, and what will be verified at the macro level.
- A specification of the target code coverage for each subblock, and for the top level macro.
- A description of the regression test environment and regression procedure. The regression tests are those verification tests that are routinely run to verify that the design team has not broken existing functionality while adding new functionality.
- A results verification procedure, specifying what criteria will be used to determine when the verification process has successfully been completed.

7.1.2   Verification Strategy

The verification of a macro consists of three major phases:

- Verification of individual subblocks
- Macro verification
- Prototyping

The overall verification strategy is to achieve a very high level of test coverage at the subblock level, and then to focus the macro-level verification at the interfaces between blocks, overall macro functionality, and corner cases of behavior. Realistically, this approach gains high but not 100 percent confidence in the macro's functional correctness. Building a rapid prototype of the macro allows the team to run significant amounts real application code on the macro, greatly increasing confidence in the robustness of the design.

At each phase of the verification process, the team needs to decide what kinds of tests to run, and what verification tools to use to run them.

The basic types of verification tests include:

Compliance testing

These tests verify that the design complies with the specification. For an industry standard design, like a PCI interface or an IEEE 1394 interface, these tests also verify compliance to the published specification. In all cases, compliance to the functional specification for the design is checked as fully as possible.

Corner testing

These tests try to find the complex scenarios, or corner cases, that are most likely to break the design. They focus on the aspects of the design that are most complex, involve the most interaction between blocks, or are the least clearly specified.

Random testing

For many designs, such as processors or complex bus interfaces, random tests can be a useful complement to compliance and corner testing. Focused tests like the compliance and corner tests are limited to the scenarios that the engineers anticipate. Random tests can create scenarios that the engineers do not anticipate, and often uncover the most obscure bugs in a design.

Real code testing

One of the most important parts of verifying any design is running the design in a real application, with real code. It is always possible for the hardware design team to misunderstand a specification, and design and test their code to an erroneous specification. Running the real application code is a useful way to uncover these errors.

Regression testing

As tests are developed, they should be added to the regression test suite. This regression test suite can then be run on a regular basis during the verification phase of the project. One of the typical problems in verification is that, in the process of fixing a bug, another bug can be inadvertently introduced. The regression test suite can help verify that the existing baseline of functionality is maintained as new features are added and bugs are fixed. It is particularly important that, whenever a bug is detected, the test case for the bug is added to the regression suite.

The verification tools available to the macro design team include:

Simulation

Most of the macro verification is performed by simulating the RTL on an event-driven simulator. Event-driven simulators give a good compromise between fast compile times and good simulation speed at the RTL level, and provide a good debug environment. For large macros, the run-time for simulation may become a problem, especially for regression tests, random tests, and real code testing. In this case, it may be worthwhile to use a cycle-based simulator, which can provide improved runtime performance, but does not provide as convenient a debug environment.

Although most simulation should be done at the RTL level, some simulation should be run at the gate level. Typically this is done late in the design cycle, once the RTL is stable and well-verified. Some initialization problems are masked at the RTL level, since RTL simulation uses a more abstract model for registers, and thus does not propagate X's as accurately as gate-level simulation. Usually only the reset sequence and a few basic functional tests need to be run at the gate level to verify correct initialization.

Hardware modeling

A hardware modeler provides an interface between a physical chip and the software simulator, so that stimulus can be applied to the chip and responses monitored within the simulation environment. Hardware modelers allow the designer to compare the simulation results of the RTL design with those of an actual chip. This verification method is very effective for designs where there is a known-good chip whose functionality is being designed into the macro.

Emulation

Emulation provides very fast run times but long compile times. It is significantly more expensive and more difficult to use than simulation. It is an appropriate tool for running real code on a large design, but is not a very useful tool for small macro development.

Prototyping

Building an actual prototype chip using the macro is key to verifying functionality. A prototype allows execution of real code in a real application at real-time speeds. A physical chip is not as easy to debug as a simulation of the design, so prototyping should only occur late in the design phase. Once a problem is detected using a prototype, it is usually best to recreate the problem in the simulation environment, and perform the debug there.

7.1.3   Subblock Simulation

Subblock verification is generally performed by the creator of the subblock, using a handcrafted testbench. This testbench typically consists of algorithmically-generated stimuli and a monitor to check outputs. The goal at this stage is 100 percent statement and path coverage, as measured with a commercial code coverage tool. This level of coverage is usually achievable with a reasonable effort because the subblocks are small. It is essential that this level of coverage be achieved at the subblock level because high levels of coverage become increasingly more difficult at higher levels of integration. Of course, good judgement needs to be used when applying this guideline. For example, if a datapath and its control block are initially designed as separate subblocks, then it may be impossible to get high coverage testing them separately. It may be much more appropriate to integrate the two and then perform verification.

Whenever possible, the outputs of the subblock should be checked automatically. The best way to do this is to add checking logic to the testbench. Of course, the checking logic needs to be even more robust than the macro code it is checking.

Automated response checking is superior to visual verification of waveforms because:

- It is less error-prone.
- It enables checking of longer tests.
- It enables checking of random tests.

Guideline – All response checking should be done automatically. It is not acceptable for the designer to view waveforms and determine whether they are correct.

Guideline – All subblock test suites should achieve 100 percent statement and path coverage as measured by a test coverage tool such as VeriSure or VHDLCover. Subblock testing is the easiest place to detect design errors. With high coverage at this level, integration-level errors should be limited to interfacing problems.

7.1.4   Macro Simulation

If the subblocks have been rigorously tested, then the major source of errors at the macro integration level will either be interface problems or the result of the designer misunderstanding the specification. Macro-level tests focus on these areas. At the macro level, 100 percent coverage is no longer a practical goal. The emphasis at this stage is on testing the interaction between the component subblocks and the interfaces of the macro with its environment. Testing random cases of inputs and outputs is a crucial element of macro verification.

The design of testbenches for macro simulation is discussed in the next section, "Testbench Design."

7.1.5   Prototyping

The design reuse methodology encourages rapid prototyping to complement simulation, and to compensate for the less-than-100 percent coverage at the macro verification stage. Achieving the final small percent of coverage at the macro level is generally extremely costly in time and still does not detect some of the bugs that will become apparent in prototype operation.

For many macros, it is possible to build a prototype chip and board and thus test the design in the actual target environment. Current FPGA and laser prototyping technologies do not provide the gate-count or the speed of state-of-the-art ASIC technology, but do provide the ability to create prototypes very rapidly. For designs that fit in these technologies and that can be verified at the speeds they provide, these technologies are very useful debugging mechanisms.

Building a prototype ASIC is required for macros that must be tested at speeds or gate counts exceeding those of FPGA and laser technologies. For some projects, this may mean that the prototype chip for the macro is the first SoC design in which it is used. In this case, the team must realize that the chip is a prototype, and that there is a high likelihood that it will have to be turned in order to achieve fully functional silicon.

7.1.6   Limited Production

Even after robust verification and prototyping, we cannot be sure that there are no remaining bugs in the design. There may be testcases that we did not run, or configurations that we did not prototype. Fundamentally, we have done a robust design but we have not used the macro in a real SoC design. For this reason, we recommend a period of limited production for any new macro. Typically, limited production involves working with just a few (1-4) customers and making sure that they are successful using the macro before releasing the macro to widespread distribution. We have found this cautious approach very beneficial in reducing the risk of support problems.

7.2   Testbench Design

Testbench design differs depending on the function of the macro. For example, the top-level testbench for a microprocessor macro would typically execute test programs, while that of a bus interface macro would typically use bus functional models and bus monitors to apply stimulus and check the results. There are also significant differences between the subblock testbench design and top-level macro testbench design. In all cases, it is important to make sure that the test coverage provided by the testbench is adequate.

7.2.1   Subblock Testbench

The testbenches for subblocks tend to be rather ad hoc, developed specifically for the subblock under test. At some abstract level, though, they tend to look like Figure 7-1.

Because subblocks will almost never have bidirectional interfaces, we can develop a simple testbench that generates a set of inputs to the input ports and checks the outputs at the output ports. The activity at these ports is not random; in most digital systems, there will be a limited set of *transactions* that occur on a given port. These transactions usually have to do with reading or writing data to some storage element (registers, FIFOs, or memories) in the block.

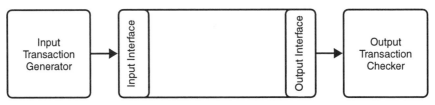

Figure 7-1   Typical testbench for a subblock

Stimulus Generation

When we design the subblock, we can specify the transaction types that are allowed to occur on a given input port; for instance, a register write consists of one specific sequence of data, address, and control pins changing, and no other sequence of actions on these pins is legal. As we design the macro, of course, we need to make sure that no block driving this port can ever generate any transactions other than the legal transaction at this port.

Once we have defined the legal set of transaction types on the input ports, we need to generate sequences of these transactions with the appropriate data/address values for testing the subblock. We start by analyzing the functionality of the subblock to determine useful sequences that will verify that the subblock complies with the specification. Then we search for the corner cases of the design: those unique sequences or combinations of transactions and data values that are most likely to break the design.

Once we have developed all the tests we can in this manner, we run a code coverage tool. This tool gives us a good indication of the completeness of the test suite. If additional testing is required to achieve 100% coverage, then we can develop additional focused tests or we can create a random test generator to generate random patterns of transactions and data. Random testing is effective for processors and bus interfaces because of the large number of transaction types make it difficult to manually generate all of the interesting combinations of transactions.

Output Checking

Generating test cases is, of course, just the first part of verification. We must check the responses of the design to verify that it is working correctly. This checking can be done manually, by monitoring the outputs on a waveform viewer, and verifying the waveforms of the outputs. This process is very error-prone, and so an automatic output checker is a necessary part of the testbench. The design of this checker is unique to the subblock being tested, but there are some common aspects to most checkers:

- We can verify that only legal transactions are generated at the output port of the design. For instance, if the read/write line is always supposed to transition one clock cycle before data and be stable until one clock cycle after data transitions, then we can check this automatically.

- We can verify that the specific transactions are correct responses to the input transactions generated. This requires a detailed analysis of the design. Clearly, the simpler the design, the simpler this checking is. This provides another reason for keeping the design as simple as possible and still meet function and performance goals.

7.2.2 Macro Testbench

We can extend the concepts used in the subblock testbench to the testbench used for checking the macro. Once the subblocks have integrated into the macro, we construct a testbench that again automatically generates transactions at the macro input ports and checks transactions at the output ports. But there are several reasons that we want to develop a more powerful and well-documented testbench at this level:

- The design is now considerably more complex, and so more complex test scenarios will be required for complete testing.

- More people will typically be working on verification of the macro, often the entire team that developed the subblocks.

- The testbench will be shipped along with the macro so that the customer can verify the macro in the system design.

The testbench can take several forms. An interface macro, such as a PCI interface, might have a testbench like the one shown in Figure 7-2.

In this testbench, a PCI bus functional model is used to create transactions on the PCI bus, and thus to the PCI macro. A PCI bus monitor checks the transactions on the PCI bus, and thus acts as a transaction checker. Because the PCI macro acts as a bridge between the PCI bus and the application bus, we need a bus functional model and a bus monitor for the application bus as well. When this testbench is fully assembled, we have the ability to generate and check any sequence of transactions at any port of the PCI macro.

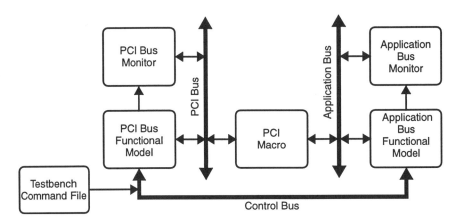

Figure 7-2   Macro development and verification environment

The testbench design is made much easier if the bus functional models and bus monitors can be controlled by a single testbench command file, so that complex, well-coordinated test scenarios can be constructed. Figure 7-2 shows a common control bus that coordinates the various models and monitors.

A more complex testbench is shown in Figure 7-3. Here, the actual software application is the source of commands for the PCI bus functional model. This application can run on the workstation that is running the simulator; device driver calls that would normally go to the system bus are redirected through a translator to the simulator, using a programming language interface such as Verilog's PLI or ModelSim's FLI. A hardware/software cosimulation environment can provide an effective way of setting up this testbench and a convenient debug environment.

The actual transactions between the application and the PCI macro under test are a small percentage of the cycles being simulated; many cycles are spent generating inputs to the bus functional model. Also, real code tends repeat many of the same basic operations many times; extensive testing of the macro requires the execution of a considerable amount of application code. Thus, software-driven simulation is an inherently inefficient test method, but it does give the opportunity of testing the macro with real code. For large macros, this form of testing is most effective if the simulation is running on a very high-speed simulator, such as a cycle-based simulator or an emulator.

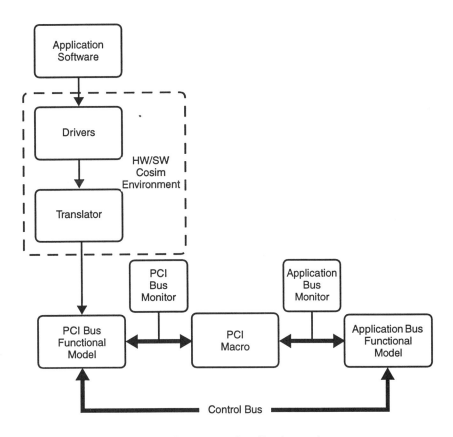

Macro development and verification environment

Figure 7-3   Software-driven testbench for macro-level testing

7.2.3   Bus Functional Models

The bus functional models (BFM) used in the examples above are a common method of creating testbenches. Typically they are written in RTL and use some form of command language to create sequences of transactions on the bus. The intent of these models is to model only the bus transactions of an agent on the bus. They do not model any of the functionality of an agent on the bus; each read and write transaction is specified by the test developer explicitly.

Because of their simplicity, these models place little demand on simulator performance; simulation speeds are mostly determined by the macro itself.

Well-designed BFMs allow the test developer to specify the transactions on the bus at a relatively high level of abstraction. Instead of controlling individual signals on the bus, the test developer can specify a read or write transaction, with the associated data, or an interrupt. The developer may well also want to generate an error condition, such as forcing a parity error; therefore, the BFM should include this capability as well.

Many testbenches require multiple BFMs. In this case, it is best to use a single command file to coordinate the actions of the various models. The models must be written so that they can share a common command file. Many commercial BFMs offer this capability. Figure 7-4 shows an example of the kind of complex test scenario that can be created with multiple BFMs. Two PCI BFMs are configured as bus masters and two are configured as bus slaves. Simultaneous, or nearly simultaneous, bus accesses can be generated, forcing the PCI macro to deal with interrupted transactions.

BFMs are also extremely useful for system-level simulation, as described in Chapter 11. For instance, a PCI BFM can be used to generate transactions to a SoC design that has a PCI interface block. For this reason, the BFM is considered one of the macro deliverables.

Because it is one of the deliverables that will ship with the product, the BFM must be designed and coded with the same care as the macro RTL. The BFM designer also needs to provide full documentation on how to use the BFM.

7.2.4   Automated Response Checking

In the previous examples, the automated response checking for the testbench was provided by the bus monitors. This approach is useful for bus interfaces, but for other types of macros there are some other techniques that may be useful.

One effective technique is to compare the output responses of the macro to those of a reference design. If the macro is being designed to be compatible with an existing chip, say, a microcontroller or DSP, then the chip itself can be used as a reference model. A hardware modeler can be used to integrate the physical chip as a model in the simulation environment. Figure 7-5 shows a such a configuration.

If a behavioral model for the design was developed as part of the specification process, then this behavioral can be used as the reference model, especially if it is cycle accurate.

One approach often used by microprocessor developers is to develop an Instruction Set Architecture (ISA) model of the processor, usually in C. This ISA model is defined to be the reference for the design, and all other representations of the design must exhibit the same instruction-level behavior. As RTL is developed for the design, its behavior is constantly being compared to the reference ISA model.

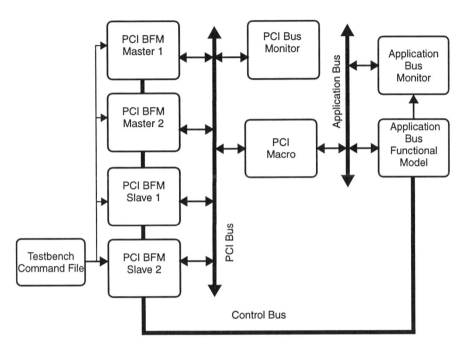

Figure 7-4   Multi-BFM PCI testbench

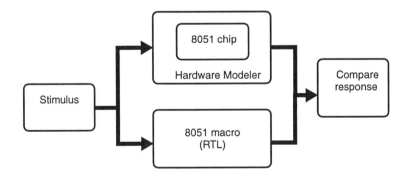

Figure 7-5   Self-checking testbench using a hardware modeler

7.2.5   Code Coverage Analysis

Verifying test coverage is essential to the verification strategy; it is the only way to assess quantitatively the robustness of the test suite. Several commercial tools are available that provide extensive coverage capabilities. We describe here some of the capabilities of the TransEDA VHDLCover tool, which is representative of the better coverage tools currently available.

The coverage tool provides the following metrics:

- Statement coverage
- Branch coverage
- Condition coverage
- Path coverage
- Toggle coverage
- Triggering coverage

Statement coverage gives a count, for each executable statement, of how many times it was executed.

Branch coverage verifies that each branch in an if-then-else or case statement was executed.

Condition coverage verifies that all branch sub-conditions have triggered the condition branch. In Example 7-1, condition coverage means checking that the first line was executed with $a = 1$ and that it was executed with $b = 0$, and it gives a count of how many times each condition occurred.

Example 7-1   Condition coverage checks branch condition

```
if (a = '1' or b = '0') then
  c <= '1';
else
  c <= 0;
endif;
```

Path coverage checks which paths are taken between adjacent blocks of conditional code. For instance, if there are two successive if-then-else statements, as in Example 7-2, path coverage checks the various combinations of conditions between the pair of statements.

Example 7-2 Path coverage

```
if (a = '1' or b = '0') then
  c <= '1';
else
  c <= 0;
endif;

if (a = '1' and b = '1') then
  d <= '1';
else
  d <= 0;
endif;
```

There are several paths through this pair of if-then-else blocks, depending on the values of *a* and *b*. Path coverage counts how many times each possible path was executed.

Triggering coverage checks which signals in a sensitivity list trigger a process.

Trigger coverage counts how many times the process was activated by each signal in the sensitivity list changing value. In Example 7-3, trigger coverage counts how many times the process is activated by signal *a* changing value, by signal *b* changing value, and by signal *c* changing value.

Example 7-3 Trigger coverage

```
process (a, b, c)



. . . . .
```

Toggle coverage counts how many times a particular signal transitions from '0' to '1', and how many times it transitions from '1' to '0'.

Achieving high code coverage with the macro testbench is a necessary but not sufficient condition for verifying the functionality of the macro. Code coverage does nothing to verify that the original intent of the specification was executed correctly. It also does not verify that the simulation results were ever compared or checked. Code coverage only indicates whether the code was exercised by the verification suite.

On the other hand, if the code coverage tool indicates that a line or path through the code was not executed, then clearly the verification suite is not testing that piece of code.

We recommend targeting 100% statement, branch and condition coverage. Anything substantially below this number may indicate significant functionality that is not being tested.

Path, toggle, and triggering coverage can be used as a secondary metric. Achieving very high coverage here is valuable, but may not be practical. At times it may be best to examine carefully sections of code that do not have 100% path, toggle, or trigger coverage, to understand why the coverage was low and whether it is possible and appropriate to generate additional tests to increase coverage.

One of the limitations of current code coverage tools is in the area of path coverage. Path coverage is usually limited to adjacent blocks of code. If the design has multiple, interacting state machines, this adjacency limitation means that it is unlikely that the full interactions of the state machines are checked.

7.3   RTL Testbench Style Guide

Behavioral testbenches work well with event-driven simulators, but they slow down simulation considerably when used with cycle-based simulators or emulation. Both cycle-based simulation and emulation work best with synthesizable RTL. For cycle-based simulators, behavioral code must be cosimulated on an event-driven simulator. For emulators, behavioral code must be cosimulated on the host workstations.

Thus, for optimal verification speed, testbenches should be coded as much as possible in the same synthesizable subset used for the macro itself. Exceptions should be isolated in separate modules, so that the smallest possible parts of the testbenches are cosimulated.

7.3.1   General Guidelines

Guideline – Partition testbench code into synthesizable and non-synthesizable (behavioral) sections. Use behavioral code to generate clocks and resets and use synthesizable code to model a finite state machine that manipulates and generates stimuli for the design, as shown in Figure 7-6.

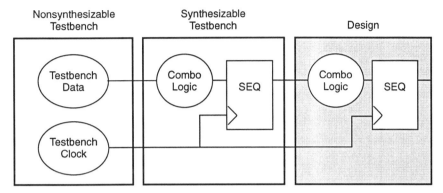

Figure 7-6   Keeping synthesizable and non-synthesizable processes separate

7.3.2   Generating Clocks and Resets

Guideline – Do not generate data, clocks, and resets from the same process. Use separate processes for clock generation, data generation, and reset generation, as shown in Figure 7-7.

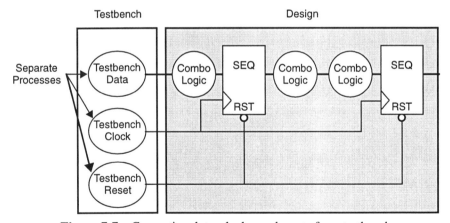

Figure 7-7   Generating data, clocks, and resets from testbench

Guideline – If you have multiple asynchronous clocks, use a separate process for each clock generation as shown in Figure 7-8.

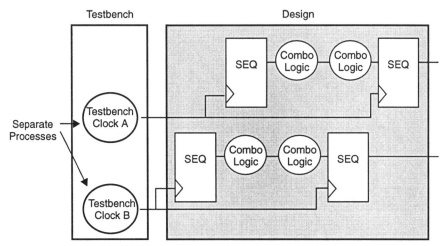

Figure 7-8   Generating multiple clocks from the testbench

Guideline (VHDL only) – Use built-in textio packages instead of user-defined packages for reading stimulus files. See Example 7-4.

Example 7-4   Using the IEEE textio package to read std_logic types

Poor coding style:

```
use std.textio.all;
  procedure read_hex(l:inout line; value: out
      std_logic_vector) is
  begin
    bit_pos:=1023;
    while (l'length>0) loop
      read(l,chr);
      if (chr/=lf or chr/=nul) then
        case chr is
          when '0' => hex_char:="0000";
          when '1' => hex_char:="0001";
          when '2' => hex_char:="0010";
            . . .
begin -- architecture
read_it:    process
  variable vline : line;
  begin
    readline (infile, vline);
    read_hex (vline, v1);
```

Recommended coding style:

```
use IEEE.std_logic_textio.all;
begin -- architecture
read_it:   process
  variable vline : line;
  begin
  readline (infile, vline);
  hread (vline, v1);
```

Guideline – Read and apply one vector per clock. Do not advance simulation time based on time value read in from the text file (Example 7-5). Reading the time value from the stimulus file creates a much more complex timing dependency between stimulus generation and the (fully synchronous) macro, slowing down cycle-based simulation and emulation.

Example 7-5   Reading and applying vectors

Poor coding style:

```
junk: process
begin
  while not endfile( infile ) loop
    readline( infile, L );
    read(L, vNEWTIME);
    wait for vNEWTIME - $NOW;
    read(L, vCLK);
    CLK <= vCLK;
    read(L, vDATA);
    DATA <= vDATA;
    ...
end loop;
```

Recommended coding style:

```
clock: process begin
  CLK <= '0';
  wait for 25 ns;
  CLK <= '1';
  wait for 25 ns;
end process;
data: process begin
  while not endfile(infile) loop
    readline(infile, L);
    wait for 50 ns; -- One clock edge
```

```
        read(L, vDATA);
        DATA <= vDATA;
   end process;
```

Guideline – Use a clock to synchronize stimulus generation. Apply all data at once on cycle boundaries and use as few individual process waits as possible.

7.4 Timing Verification

Static timing verification is the most effective method of verifying a macro's timing performance. As part of the overall verification strategy for a macro, the macro should be synthesized using a number of representative library technologies. Static timing analysis should then be performed on the resulting netlists to verify that they meet the macro's timing objectives.

The choice of which libraries to use is a key one. Libraries, even for the same technology (say .5µ), can have significantly different performance characteristics. The libraries should be chosen to reflect the actual range of technologies in which the macro is likely to be implemented.

For macros that have aggressive performance goals, it is necessary to include a trial layout of the macro to verify timing. Pre-layout wireload models are statistical and actual wire delays after layout may vary significantly from these models. Doing an actual layout of the macro can raise the confidence in its abilities to meet timing.

Gate-level simulation is of limited use in timing verification. While leading gate-level simulators have the capacity to handle 500k or larger designs, gate-level simulation is slow. The limited number of vectors that can be run on a gate-level netlist cannot exercise all of the timing paths in the design, so it is possible that the worst case timing path in the design will never be exercised. For this reason, gate-level timing simulation may deliver optimistic results and is not, by itself, sufficient as a timing verification methodology.

Gate-level simulation is most useful in verifying timing for asynchronous logic. We recommend avoiding asynchronous logic, because it is harder to design correctly, to verify functionality and timing, and to make portable across technologies and applications. However, some designs may require a small amount of asynchronous logic. The amount of gate-level, full timing simulation should be tailored to the requirements of verifying the timing of this asynchronous logic.

Static timing verification, on the other hand, tends to be pessimistic unless false paths are manually defined and not considered in the analysis. Because this is a manual process, it is subject to human error. Gate-level timing simulation does provide a coarse check for this kind of error.

The best overall timing verification methodology is to use static timing analysis as the basis for timing verification. You can then use gate-level simulation as a second-level check for your static timing analysis methodology (for example, to detect mis-identified false paths).

CHAPTER 8	*Developing Hard Macros*

This chapter discusses issues that are specific to the development of hard macros. In particular, it discusses the need for simulation, layout, and timing models, as well as the differing productization requirements and deliverables for hard macros. The topics are:

- Overview
- Hard macro design issues
- Hard macro design process
- Physical design for hard macros
- Block integration
- Productization
- Model development for hard macros
- Portable hard macros and CBA technology

8.1 Overview

Hard macros are macros that have a physical representation, and are delivered in the form of a GDSII file. As a result, hard macros are more predictable than soft macros in terms of timing, power, and area. But hard macros do not have the flexibility of soft macros; they cannot be parameterized or user-configurable. The porting process of the two forms is also quite different; hard macros must be ported at the physical level, while soft macros merely need to be re-synthesized to the new library.

In this document, we assume that the responsibility for the physical integration of the macro belongs to the silicon vendor. This is the most common model today for hard macros: the silicon vendor owns the hard macro and does not make the GDSII available to the system integrator. The system integrator uses a set of models of the hard macro to design the rest of the chip, and the silicon vendor does the final place and route, integrating the hard macro GDSII into the final design.

In some cases, of course, the silicon vendor is also the system designer; but the processes described in this chapter are still applicable. Exceptions will be noted in the appropriate sections of this chapter.

Hard macro development is essentially identical to soft macro development except for two major issues:

- The design process for hard macros includes the added activity of generating a physical design.
- Hard macros require models for simulation, layout, and timing.

These requirements stem from the fact that hard macros are delivered as a physical database rather than RTL. Integrators require these models to perform system-level verification, chip-level timing, floorplanning, and layout.

It is recommended that the design process itself be kept identical with the design process for soft macros except for the productization phase. The following sections describe how the design process for hard macros differs from the design process for soft macros.

8.2   Design Issues for Hard Macros

There are several key design issues that are unique to hard macros. These issues affect the design process, and are described in the following sections.

8.2.1   Design For Test

Hard macros pose some unique test issues not found in soft macros. With soft macros, the integrator can choose from a variety of test methodologies: full scan, logic BIST, or application of parallel vectors through boundary scan or muxing out to the pins of the chip. The actual test structures are inserted at chip integration, so that the entire chip can have a consistent set of test structures.

Hard macros do not provide this flexibility; test structures must be built into each hard macro. The integrator then must integrate the test strategy of the hard macro with the test strategy for the rest of the chip. It is the task of the hard macro developer to provide an appropriate test structure for the hard macro that will be easy to integrate into a variety of chip-level test structures.

The hard macro developer must choose between full scan, logic BIST, or application of parallel vectors through boundary scan or muxing out to the pins of the chip.

Full scan offers very high test coverage and is easy to use. Tools can be used to insert scan flops and perform automatic test pattern generation. Fault simulation can be used to verify coverage. Thus, scan is the preferred test methodology for hard macros as long as the delay and area penalties are acceptable. For most designs, the slight increase in area and the very slight increase in delay are more than compensated by the ease of use and robustness of scan.

For some performance-critical designs, such as a microprocessor, a "near full scan" approach is used, where the entire macro is full scan except for the datapath, where the delay would be most costly. For the datapath, only the first and last levels of flops are scanned.

Logic BIST is a variation on the full scan approach. Where full scan must have its scan chain integrated into the chip's overall scan chain(s), logic BIST uses an LFSR (Linear Feedback Shift Register) to generate the test patterns locally. A signature recognition circuit checks the results of the scan test to verify correct behavior of the circuit.

Logic BIST has the advantage of keeping all pattern generation and checking within the macro. This provides some element of additional security against reverse engineering of the macro. It also reduces the requirements for scan memory in the tester. Logic BIST does require some additional design effort and some increase in die area for the generator and checker, although tools to automate this process are becoming available.

Parallel vectors are used to test only the most timing or area critical designs. A robust set of parallel vectors is extremely time-consuming to develop and verify. If the macro developer selects parallel vector testing for the macro, boundary scan must be included as part of the macro. Boundary scan provides an effective, if slow, way of applying the vectors to the macro without requiring muxing the macro pins out to the chip pins. Requiring the integrator to mux out the pins places an unreasonable burden on the integrator and restricts the overall chip design.

8.2.2   Clock and Reset

The hard macro designer has to implement a clock and reset structure in the hard macro without knowing in advance the clocking and reset structure of the chip in which the macro will be used. The designer should provide full clock and reset buffering in the hard macro, and provide a minimal load on the clock and reset inputs to the macro.

To ease the integration of the macro onto the chip, the designer should provide a buffered, precisely aligned copy of the clock as an output of the macro. This clock is then available to synchronize the rest of the chip's clocking.

8.2.3   Aspect Ratio

The aspect ratio of the hard macro affects the floorplan and routability of the final chip. Thus, it is an important factor affecting the ease with which the macro can be integrated in the final chip. A large hard macro with an extreme ratio can present significant problems in placing and routing a SoC design. In most cases, an aspect ratio close to 1:1 minimizes the burden on the integrator. Aspect ratios of 1:2 and 1:4 are also commonly used.

8.2.4   Porosity

Hard macros can present real challenges to the integrator if they completely block all routing. Some routing channels through the macro should be made available to the integrator, if it is possible to do so without affecting the macro's performance.

Another approach is to limit the number of used metal layers to less than the total available in the process. For processes with more than two metal layers available for signal routing, this can be an effective approach to providing routing through the hard macro.

At the very least, the macro deliverables should include a blockage map to identify areas where over-cell routing will not cause errors due to crosstalk or other forms of interaction.

8.2.5   Pin Placement

Pin placement and ordering can also affect the floorplan of the chip. A floorplanning model is one of the deliverables of a hard macro. Among other things this model describes the pin placement, size, and grid.

8.2.6   Power Distribution

Power and ground busing within the macro must be designed to handle the peak current requirements of the macro at maximum frequency. The integrator using the macro must provide sufficient power busing to the macro to limit voltage drop, noise, and simultaneous output switching noise to acceptable levels. The specification of the hard macro must include sufficient information about the requirements of the macro and the electrical characteristics of the power pin contacts on the macro.

8.3   The Hard Macro Design Process

The hard macro design process is shown in Figure 8-1. For the hard macro, we expand the macro specification to include physical design issues. The target library is specified, and timing, area, and power goals are described.

The macro specification also addresses the issues described in the previous section: design for test, clock and reset, aspect ratio, porosity, pin placement, and power distribution. The specification describes the basic requirements for each of these.

The macro specification also describes the models that will provided as part of the final deliverables. These models include the simulation model(s), timing model(s), and floorplanning model.

Concurrent with the functional specification and behavioral model development, we develop a more detailed physical specification for the macro, addressing all of the issues mentioned above, describing how each requirement of the macro specification will be met. From this specification, we develop a preliminary floorplan of the macro. This floorplan and the physical requirements of the macro help drive the partitioning of the macro into subblocks.

Once the macro is partitioned into subblocks, the design of the individual subblocks follows the same process as for soft macros.

For some very high performance designs, the designer may elect to not to use automated synthesis for some critical subblocks. Instead, the designer may use a datapath compiler or may hand craft the subblock. The goal of these alternate synthesis methods is the same: to meet the timing, area, and power requirements of the macro specification while ensuring that the detailed design is functionally equivalent to the RTL.

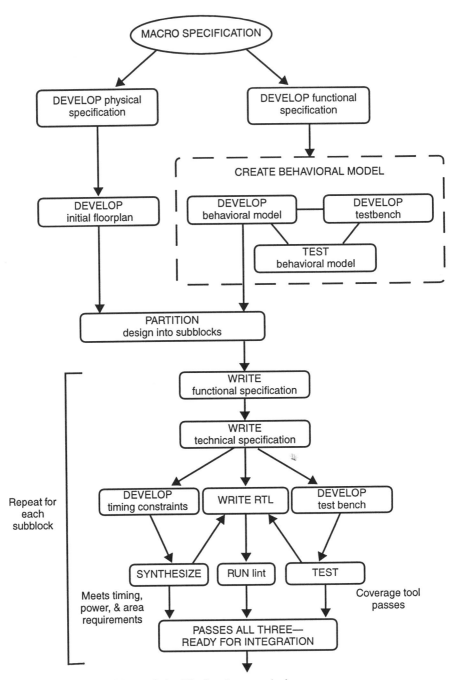

Figure 8-1   The hard macro design process

Note that even with manual synthesis and handcrafting, the RTL for the subblock is the "golden" reference. For all synthesis methods, automated and manual, formal verification should be used to ensure the equivalence between the final physical design and the RTL.

8.4   Block Integration for Hard Macros

The process of integrating the subblocks into the macro is much the same for both hard and soft macros. This process is described in Figure 8-2.

Because a hard macro is available only in a single configuration, functional test is somewhat simplified; no multiple-configuration testing is required, as it is for soft macros.

As described in the previous section, manufacturing test presents additional challenges to hard macro design. Based on the requirement for the macro, a test methodology must be selected and implemented.

Synthesis needs to target only the target technology library. Because porting is done at the physical level, after synthesis, there is no requirement to produce optimal netlists in a variety of technologies.

Synthesis of the macro is an iterative process, involving refining the floorplan based on synthesis results, updating the wireload models based on the floorplan, and repeating synthesis. With a good initial floorplan, good partitioning of the design, and good timing budgets, this process will converge rapidly. As the process starts to converge, an initial placement of the macro that produces an estimated routing can further improve the wireload models used for synthesis.

Integrating Subblocks into a Macro

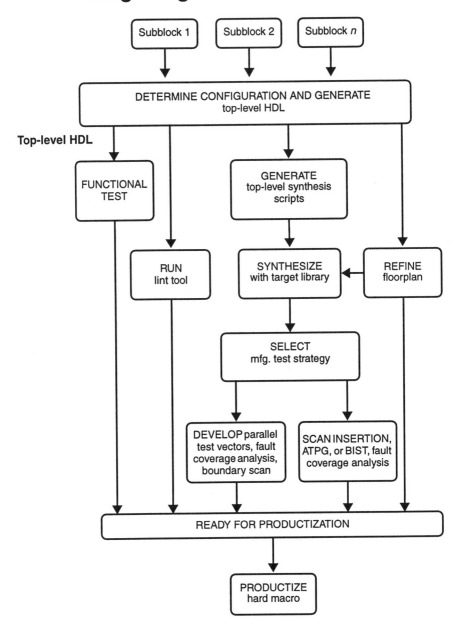

Figure 8-2 Flow for integrating subblocks into the hard macro

8.5   Productization of Hard Macros

Productization of hard macros involves physical design, verification, model development, and documentation.

8.5.1   Physical Design

The first step in productizing the hard macro is to complete the physical design. Figure 8-3 shows the basic loop of floorplanning and incremental synthesis, place and route, and extracting timing. In the first pass of this loop, the final floorplan and synthesized netlist provide the inputs to the place and route tools. After the initial place and route, actual delay values are extracted from the physical design and delivered back to the synthesis tool in the form of an SDF file. We can then perform static timing analysis to determine if the design meets our timing goals. If necessary, we can also perform power analysis to see if the design meets the power goals.

If the physical design does not meet timing, we have two choices. If timing, power, or area is far from meeting specification, we may need to go back to the design phase and iterate as required. If we are reasonably close to meeting specification, however, we can focus on synthesis only. We first try using the IPO (In Place Optimization) feature of the synthesis tool. IPO modifies as little of the design as possible, focusing on resizing buffers. We then provide the updated netlist to the place and route tool and do an incremental place and route, where only the updated gates are modified in the physical design. By retaining as much of the original place and route as possible, we optimize our chances of rapidly converging on a good place and route.

8.5.2   Verification

Once we achieve our physical design goals with a place and route, we perform a series of verifications on the physical design:

1. **Gate verification** – We use formal verification to prove that the final gate-level netlist is equivalent to the RTL. For hand-crafted blocks, we use a combination of LVS (Layout vs. Schematic), to verify transistor to gate netlist equivalence, and formal verification. We also run full-timing gate-level simulation to verify any asynchronous parts of the design.
2. **Static Timing Analysis** – We perform a final static timing analysis to verify that the design meets timing.
3. **Physical Verification** – We use LVS and DRC (Design Rule Checking) tools to verify the correctness of the final physical design.

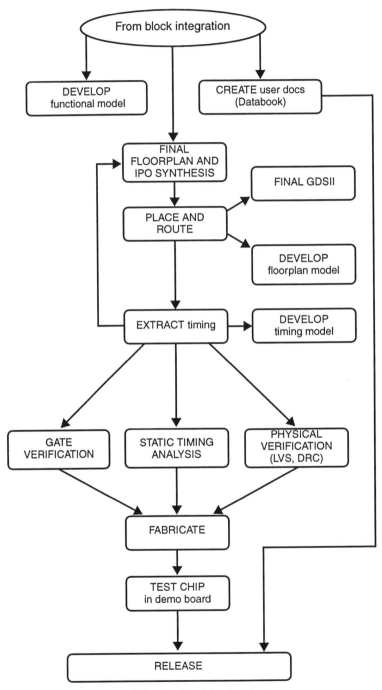

Figure 8-3   Productizing hard macros

8.5.3   Models

In addition to the physical design database, we need to develop the models that the integrator will use to model the macro in the system design:

- The functional simulation model is developed from the final RTL.
- The floorplan model is developed as part of the floorplanning process.
- The timing model is developed using the extracted timing values.

The process and tools for developing these models are discussed later in this chapter.

8.5.4   Documentation

Finally, we need to create a complete set of user documentation to guide the integrator in using these models to develop the chip design. In addition to the requirements for a soft macro, the documentation for a hard macro includes:

- Footprint and size of the macro
- Detailed timing specification
- Routing restrictions and porosity
- Power and ground interconnect guidelines
- Clock and reset timing guidelines

8.6   Model Development for Hard Macros

The set of models provided to the integrator is the key to the usability of a hard macro. For most hard macros, the desired models include:

- Behavioral or ISA (Instruction Set Architecture) model for fast simulation
- Bus functional model for assisting system-level verification
- Full functional, cycle-accurate model for accurate simulation
- Functional model for emulation
- Timing and area model for static timing analysis and synthesis
- Floorplanning model

We can minimize the additional effort to create these models by leveraging the models that are created as part of the macro development process.

8.6.1   Functional Models

Currently, the process for developing functional models is ad hoc. There are no tools or processes available today for fully automating model generation. What we describe here is a process based on existing tools currently being used by the Logic Modeling Group (LMG) of Synopsys.

Model Security

One of the critical issues in developing a modeling process is determining the level of security required for the models. All the functional models described in this section are either C (or C++) models or HDL models. These models can be shipped directly to the customer if security is not a concern. If security is a concern, then some form of protection must be used.

The form of protection used by LMG involves compiling the model and using a SWIFT interface wrapper around the model to interface the model to the simulator. By delivering object code, the designer ensures a high level of security. The SWIFT interface allows the compiled model to work with all of the major commercial simulators.

Behavioral and ISA Models

Extensive hardware/software cosimulation is critical to the success of many SoC design projects. In turn, effective hardware/software cosimulation must use very high performance models for large system blocks in order to achieve the required speed. Behavioral and ISA models provide this level of performance by abstracting out many of the implementation details of the design.

Most design teams use ISA models for modeling processors during hardware/software cosimulation. These models accurately reflect the instruction-level behavior of the processor while abstracting out implementation details. This high level of abstraction allows for very fast simulation. Many processor design teams develop a C (or C++) model of the processor as they define the processor architecture. This model is then used as a reference against which the detailed design is compared. The model is then delivered to the end user of the processor.

The SoC designer using the processor core in his design can then use the ISA model to verify his software and the rest of the system design. The hardware/software cosimulation environment provides an interface between this model and the RTL simulation of the rest of the hardware, as shown in Figure 8-4.

Behavioral models are the equivalent of ISA models for non-processor designs. Behavioral models represent the algorithmic behavior of the design at a very high level of abstraction, allowing very high speed simulation. For instance, for a design using an MPEG macro, using a behavioral model instead of an RTL model can provide orders of magnitude faster system simulation.

Behavioral models can be written in C/C++, Verilog, or VHDL, and they may be protected or unprotected. If the model is shipped as compiled C, or as Verilog or VHDL source, then the hardware/software simulation environment can use it directly. If the HDL model requires protection, then the model can be compiled and linked with the SWIFT interface.

The flow and tools for compiling the behavioral Verilog/VHDL models are shown in Figure 8-5. VMC (Verilog Model Compiler) compiles the Verilog model into a VCS-compatible object format. VFM then adds the SWIFT interface, allowing the model to work with all major simulators.

If the model is coded in VHDL, then it must first be translated to Verilog. This can be done with commercial translation tools available from a number of vendors, including InterHDL. These translation tools are not yet perfected, especially for behavioral code. But they provide an initial translation that can be completed manually.

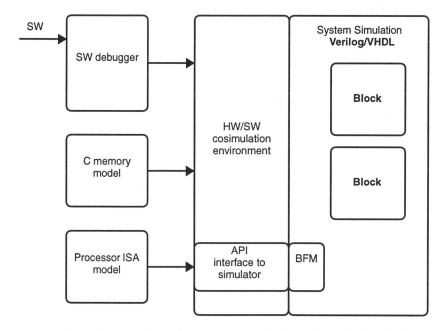

Figure 8-4   Hardware/software cosimulation using an ISA model

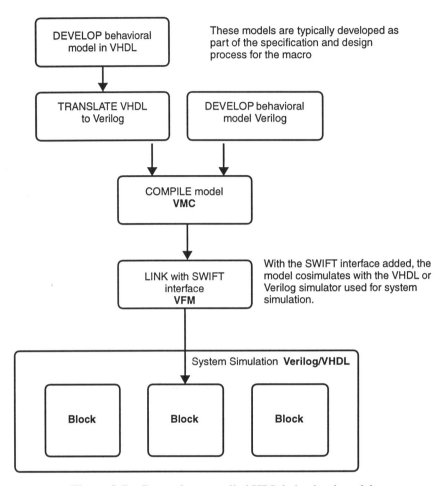

Figure 8-5   Generating compiled HDL behavioral models

Bus Functional Models

Bus functional models abstract out all the internal behavior of the macro, and only provide the capability of creating transactions on the output buses of the macro. These models are useful for system simulation when the integrator wants to test the rest of the system, independent of the macro. By abstracting out the internal behavior of the macro, we can develop a very fast model that still accurately models the detailed behavior of the macro at its the interfaces.

Bus functional models are usually developed in Verilog or VHDL, and distributed in source code. Because so little of the detailed behavior of the macro is modeled, security is not a major concern.

Full Functional Models

Although more abstract models are useful for system level verification, final verification of the RTL must be done using full functional models for all blocks in the design. Full functional models provide the detailed, cycle-by-cycle behavior of the macro. The RTL for the macro is a full functional model, and is the easiest full functional model to deliver to the integrator. Because the model is available in RTL, the flow shown in Figure 8-6 can be used. This flow is essentially the same as that for behavioral models coded in Verilog or VHDL.

Because the RTL for the macro is synthesizable, the requirement to translate VHDL to Verilog is much less of a problem than for behavioral models. Commercial translators do a reasonably good job of this translation.

The major problem with full functional models is that they are slow to simulate.

Some integrators insist on doing final system simulation with a full functional model that contains detailed timing information. This can be particularly useful in designs that exhibit asynchronous behavior, either in the hard macro or in other blocks in the system. Asynchronous design is not recommended because it is much harder to verify.

We can develop a full functional, full timing simulation model from the back-annotated netlist obtained from place and route. The same compilation scheme shown in Figure 8-6 can be used. The drawback of this approach is that simulation is extremely slow.

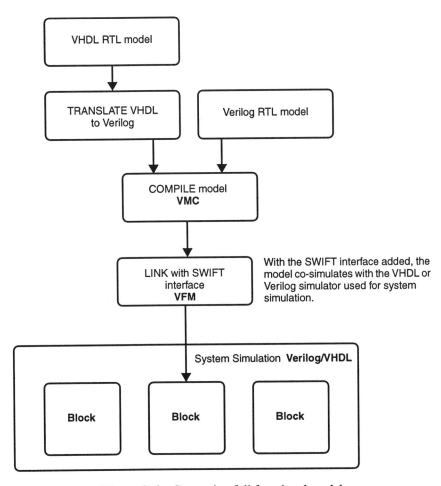

Figure 8-6   Generating full functional models

Figure 8-7 shows how to develop a full functional, full timing model with much better simulation performance. This approach takes the full functional SWIFT model developed from the RTL and adds a timing wrapper; that is, a set of structures on the inputs and outputs that can be used to model the actual delays (and setup and hold requirements) of the macro. The timing information for these buffers can be derived from the extracted timing information from place and route. This approach can be very effective provided that the macro is designed so that it does not have state dependent timing.

State dependent timing occurs when the timing characteristics of the block depend on the value of the inputs or on the internal state of the block. For instance, asynchronous RAMs have different timing for read and write modes. On the other hand, synchronous RAMs have exactly the same timing regardless of mode, and thus are easier to characterize. Using a fully synchronous design style ensures that the macro will have no state dependent timing.

It can be extremely burdensome to develop timing shells for blocks with state dependent timing, to the point where this approach is not practical.

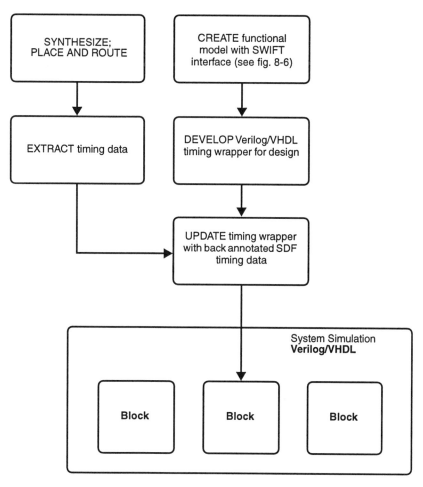

Figure 8-7 Generating full functional models with timing

Emulation Models

One of the major problems with full functional models is the slow simulation speeds achieved with them. Emulation is one approach to addressing the problem of slow system-level simulation with full functional models.

Emulation requires that the model for the macro be compiled into a gate-level representation. We can provide the RTL directly to the integrator, who can then use the emulator's compiler to generate the netlist, but this does not provide any security.

An alternate approach is to provide a netlist to the integrator. This approach provides some additional security for the macro. A separate synthesis of the macro, compiling for area with no timing constraints, will give a reasonable netlist for emulation without providing a netlist that meets the full performance of the macro.

Some emulation systems have more sophisticated approaches to providing security for hard macro models. See Chapter 11 for a brief discussion on this subject.

Hardware Models

Hardware models provide an alternate approach for providing highly secure full functional models. Because the hard macro design process requires that we produce a working test chip for the macro, this approach is often a practical form of model generation.

Hardware modelers are systems that allow a physical device to interface directly to a software simulator. The modeler is, in effect, a small tester that mounts the chip on a small board. When the pins of the device are driven by the software simulator, the appropriate values are driven to the physical chip. Similarly, when the outputs of the chip change, these changes are propagated to the software simulator.

Rapid prototyping systems, such as those from Aptix, also allow a physical chip to be used in modeling the overall system. These systems are described in Chapter 11.

Some emulators, including those from Mentor Graphics, allow physical chips to be used to model part of the system. Thus, the test chip itself is an important full functional model for the macro.

In all these cases, it is important that the physical chip reflect exactly the functionality of the macro. For example, with a microprocessor, one might be tempted to make the data bus bi-directional on the chip, to save pins, even though the macro uses unidirectional data buses. This approach makes it much more difficult to control the core and verify system functionality with a hardware modeler or emulator.

8.6.2   Synthesis and Floorplanning Models

The timing, area, and floorplanning models can be generated from the design database.

From the final place and route of the macro, we can extract the basic blockage information, pin locations, and pin layers of the macro. This information can then used by the integrator when floorplanning the SoC design. This information is typically delivered in the LEF format.

Figure 8-8 shows the process for developing a static timing analysis model for the hard macro. From the SDF back-annotated netlist for the macro, the PrimeTime timing analysis tool extracts a black-box timing model for the macro. This model provides the setup and hold timing requirements for input pins and the clock-to-output delays for the output pins. This model is delivered as a Synopsys standard format .db file. During static timing analysis on the entire chip, PrimeTime uses the context information, including actual ramp rates and output loading, to adjust the timing of the hard macro model to reflect the macro's actual timing in the chip.

For this black-box model to work, of course, the design must have no state-dependent timing. For blocks that do have state-dependent timing, a gray box timing model must be used; this model retains all of the internal timing information in the design. The entire back-annotated netlist can be used as a gray-box model, but it will result in slower static timing analysis runtimes.

If the hard macro has any blocks that are hand-crafted at the transistor level, we need another approach for extracting this timing information. Figure 8-9 shows a flow for this case. After parasitic extraction, the PathMill static timing analysis tool verifies that the timing requirements for the design are met. Through a configuration file, the designer provides the input ramp rates and output loading information, as well as identification of input, output, and clock pins. When timing has been successfully verified, PathMill can generate a black box timing model for the design in Stamp format. If desired, additional characterization information can be provided to the tool, and PathMill will develop a table of timing values based on different input ramp rates and output loading. PrimeTime uses this Stamp model to develop the final timing model in the .db format.

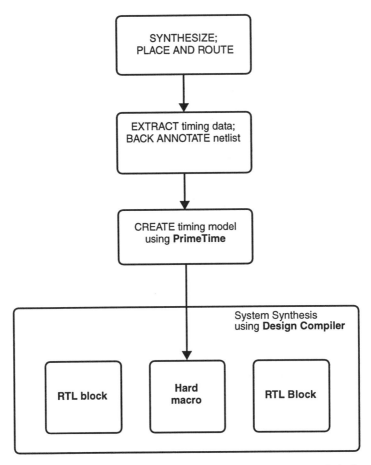

Figure 8-8   Generating static timing models for standard cell designs

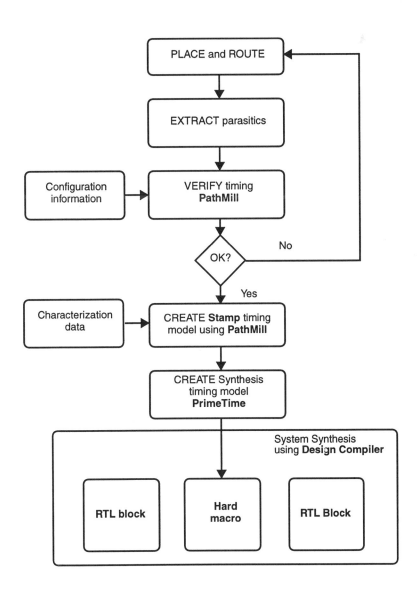

Figure 8-9   Generating static timing models for full custom designs

8.7   Portable Hard Macros and CBA

Hard macros are delivered in GDSII format; that is, a physical representation of the design is the output of the design process. Hard macros are more predictable in terms of area, power, and timing than soft macros. However, porting hard macros to different technologies is significantly more complex than porting soft macros. Porting hard macros quickly and cheaply is a major issue for hard macro providers.

Techniques for porting hard blocks include:

• Re-synthesis and re-layout of the block

• Porting by polygon manipulation

• Automated porting by retaining relative placement

The goal of the two last approaches is to reduce the effort required for the first option. Tools for automating the polygon manipulation approach are available, but they may require manual edits of the final design to get the design working.

The approach of retaining relative place can be very effective, especially if the design is fully synchronous. CBA Transport technology is one example of this approach to porting.

8.7.1   CBA Technology

The CBA (Cell-Based Array) technology is an ASIC library that exhibits many of the best characteristics of both gate array and standard cell libraries. It provides a regular tiling of cells, like a sea-of-gates gate array. However, unlike traditional gate array technology, CBA employs a tile consisting of three sizes of transistors and two types of cell: drive and compute. The drive cells employ larger transistors for driving large interconnect capacitances. The compute cells provide high speed, dense logic functions. All interconnect is provided on a regular grid of metal layers.

The multiple cell types in CBA allow designs to achieve standard cell timing performance, and near-standard cell area, in a metal-only, gate-array technology.

8.7.2   CBA Transport

The regular structure of CBA also allows for a unique IP porting capability. The CBA library has been ported to over forty different processes in twenty different foundries. As it is ported, the CBA tiling and routing geometries are maintained; that is, the relative placement of drive and compute cells, pins, power and ground connections, and all routing channels, is preserved.

As a result, a hard macro that has been designed in one CBA process can be ported to another in a fully automated fashion. Synopsys has recently announced a product, CBA Transport, that provides this capability. The basic flow for CBA Transport is shown in Figure 8-10.

There are some basic process issues that must be considered in order to take advantage of CBA Transport. The target process must have at least the same number of routing layers as the source process. The design may use stacked vias if they are supported by the target process. As long as the basic capabilities of the target process are a super-set of the source process, CBA Transport will port the IP automatically.

The reason CBA Transport is able to achieve fully automatic porting of hard IP is that the CBA library maintains the same relative physical structure as it is ported to different processes. Thus, it is able to retain the same cell placement and routing for the hard macro as it is ported from source process to target process. Power routing similarly is preserved during porting.

With any porting process, there is a chance of some relative timing changes during porting. CBA transport is most effective for synchronous designs, which are inherently less susceptible to small timing changes and easier to verify for timing.

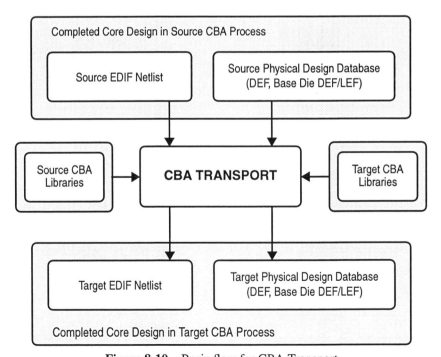

Figure 8-10   Basic flow for CBA Transport

CHAPTER 9

Macro Deployment: Packaging for Reuse

This chapter discusses macro deployment issues, including deliverables for hard and soft macros and the importance of keeping a design archive. The topics are:

- Delivering the complete product
- The contents of the user guide

9.1 Delivering the Complete Product

Once a macro has been designed according to the guidelines detailed in the preceding chapters, the macro must be packaged and delivered to the customer. The packaging of the macro depends on how it will be delivered.

Soft macros require:

- The RTL code
- Support files
- Documentation

Hard macros require:

- A rich set of models for system integration
- Documentation support for integration into the final chip

In addition, all files associated with the development of the macro must be stored together in a design archive so that all necessary information is available when it is time to modify or upgrade the macro.

As described in Chapter 4, a physical prototype of the macro is built as part of the verification phase of the macro development. Many third party vendors make this prototype available to customers either as a demonstration of capability or as an evaluation unit. Evaluation prototypes are particularly helpful with programmable macros like microcontrollers and microprocessors; application code can be developed and run on the prototype to verify functionality and performance.

9.1.1   Soft Macro Deliverables

Table 9-1 lists the deliverables for soft macros.

Table 9-1   Deliverables for soft macros

Group	Deliverables
Product files	• Synthesizable Verilog for the macro and its subblocks • Synthesizable VHDL for the macro and its subblocks • Application notes, including Verilog/VHDL design examples that instantiate the macro • Synthesis scripts and timing constraints • Scripts for scan insertion and ATPG • CBA or other reference library • Installation scripts
Verification files	• Bus functional models/monitors used in testbench • Testbench files, including representative verification tests
Documentation files	• User guide/functional specification • Data sheet
System Integration files	• Bus functional models of other system components • Cycle-based/emulation models as appropriate to the particular macro and/or its testbenches and BFMs • Recommendation of commercially available software required for hardware/software cosimulation and system integration, as appropriate for the particular macro

Product Files

In addition to the RTL in Verilog and VHDL, we must include the synthesis and installation scripts. We include the reference CBA library so that the customer can synthesize the design and verify that the installation was completed successfully. In addition, providing the reference library, scripts, and verification environment allows the user the recreate the developer's environment. This allows the user to verify many of the claims of the developer, in terms of timing, power, area, and testability of the macro.

The CBA reference library is also very helpful in identifying library problems in the integrator's environment. Synthesis libraries vary considerably. If the integrator encounters synthesis problems with the vendor's library, he can synthesize exactly the same configuration with the same scripts using the CBA library. This process helps the integrator identify whether the problem is in the macro (and its scripts) or in the vendor's technology library.

Application notes that show exactly how to instantiate the design are also useful. If the application notes are available in soft form, the integrator can cut and paste the instantiation example, avoiding typographical errors and ensuring correct port names.

Verification files

The entire verification environment, including any bus functional models, bus monitors, or other models, and some set of verification test cases are shipped with the product. The test cases that ship with the macro typically do not represent the full test suite used to verify the macro. Typically, a subset is shipped that is sufficient to ensure that the macro has been installed correctly at the integrator's site. The integrator then develops a test suite to verify the functioning of the macro in the full chip.

The bus functional models used to develop and verify the macro can be used by the integrator to create a testbench environment for the SoC chip. See Chapter 11 for more discussion on using bus functional models for system level testing.

System Integration Files

Depending on the specific macro, there may be additional deliverables that are useful for the integrator.

For large macros, where simulation speed in the system environment may be an issue, it can be useful to include cycle-based simulation and/or emulation models. In general, RTL that complies with the coding guidelines in this document will work with cycle-based simulation and emulation. However, testbenches and bus functional models, unless coded to these same RTL guidelines, may not be usable with these verification tools. It is up to the macro provider to determine which models need to be provided in cycle-based simulation/emulation compatible forms.

For macros that have significant software requirements, such as microcontrollers and processors, it is useful to include a list of compilers, debuggers, and real-time operating systems that support the macro. For other designs, we may want to reference software drivers that are compatible with the design. In most cases, the macro provider will not be providing the software itself, but should provide information on how to obtain the required software from third-party providers.

9.1.2   Hard Macro Deliverables

Table 9-2 lists the deliverables for a hard macro.

The list of deliverables in Table 9-2 assumes that the physical integration is being done by the silicon vendor rather than by the chip designer who is using the macro. This model applies when the silicon vendor is also the macro vendor. In the case where the chip designer is also doing the physical integration of the macro onto the chip, the physical GDSII design files are also part of the deliverables.

Table 9-2   Deliverables for hard macros

Group	Deliverables
Product files	• Installation scripts
Verification files	• None
Documentation files	• User guide/functional specification • Data sheet
System Integration files	• ISA or behavioral model • Bus functional model for macro • Full functional model for macro • Cycle-based simulation/emulation models as appropriate to the particular macro • Timing and synthesis model for macro • Floorplanning model for macro • Recommendation of commercially available software required for hardware/software cosimulation and system integration, as appropriate for the particular macro • Test patterns for manufacturing test, where applicable

The deliverables for hard macros consist primarily of the documentation and models needed by the integrator to design and verify the rest of the system. These models are described in Chapter 8.

For processors, an ISA model provides a high level model that models the behavior of the processor instruction-by-instruction, but without modeling all of the implementation details of the design. This model provides a high speed model for system testing, especially for hardware/software cosimulation. Many microprocessor vendors also provide a tool for estimating code size and overall performance; such a tool can help determine key memory architecture features such as cache, RAM, and ROM size.

For other macros, a behavioral model provides the high speed system-level simulation model. The behavioral model models the functionality of the macro, on a transaction-by-transaction basis, but without all the implementation details. A behavioral model is most useful for large macros, where a full-functional model is too slow for system-level verification.

For large macros, bus functional models provide the fastest simulation speed by modeling only the bus transactions of the macro. Such a model can be used to test that other blocks in the system respond correctly to the bus transactions generated by the macro.

The full functional model for the macro allows the integrator to test the full functionality of the system, and thus is key to system level verification.

As in the case of soft macros, cycle-based simulation and/or emulation models, especially for the macro testbench, may be useful for the integrator. These models are optional deliverables.

The timing and synthesis models provide the information needed by the integrator to synthesize the soft portion of the chip with the context information from the hard macro. These models provide the basic timing and loading characteristics of the macro's inputs and outputs.

The floorplanning model for macro provides information the integrator needs to develop a floorplan of the entire chip.

Test patterns for manufacturing test must be provided to the silicon manufacturer at least, if not to the end user. For scan-based designs, the ATPG patterns and control information needed to apply the test patterns must be provided. For non-scan designs, the test patterns and the information needed to apply the test patterns is required; usually access is provided through a JTAG boundary-scan ring around the macro.

9.1.3   The Design Archive

Table 9-3 lists the items that must be stored together in the design archive. All of these items are needed when any change, upgrade, or modification is made to the macro. The use of a software revision control system for archiving each version is a crucial step in the design reuse workflow, and will save vast amounts of aggravation and frustration in the future.

Table 9-3   Contents of the design archive

Group	Contents
Product files	• Synthesizable Verilog for the macro and its subblocks • Synthesizable VHDL for the macro and its subblocks • CBA reference library • Verilog /VHDL design examples that instantiate the macro • Synthesis scripts • Installation scripts
Verification files	• Bus functional models/monitors used in testbench • Testbench files
Documentation files	• User guide/functional specification • Technical specification • Data sheet • Test plan • Simulation log files • Simulation coverage reports (VHDLCover, VeriSure, or equivalent) • Synthesis results for multiple technologies • Testability report • Lint report assuring compliance to coding guidelines
System Integration files	• Bus functional models of other system components • Recommendation of commercially available software required for hardware/software cosimulation and system integration, as appropriate for the particular macro • Cycle-based simulator and hardware emulator models

9.2   Contents of the User Guide

The user guide is the key piece of documentation that guides the macro user through the selection, integration, and verification of the macro. It is essential that the user guide provides sufficient information, in sufficient detail, that a potential user can evaluate whether the macro is appropriate for the application. It must also provide all the information needed to integrate the macro into the overall chip design. The user guide should contain, at a minimum, the following information:

- Architecture and functional description
- Claims and assumptions
- Detailed description of I/O
- Exceptions to coding/design guidelines
- Block diagram
- Register manual
- Timing diagrams
- Timing specifications and performance
- Power dissipation
- Size/gate count
- Test structures, testability, and test coverage
- Configuration information and parameters
- Recommended clocking and reset strategies
- Recommended software environment, including compilers and drivers
- Recommended system verification strategy
- Recommended test strategy
- Floorplanning guidelines
- Debug strategy, including in-circuit emulation and recommended debug tools
- Version history and known bugs

The user guide is an important element of the design-for-reuse process. Use it to note all information that future consumers of your macro need to know in order to use the macro effectively. The following categories are especially important:

Claims and assumptions

Before purchasing a macro, the user must be able to evaluate its applicability to the end design. To facilitate this evaluation, the user guide must explicitly list all of the key features of the design, including timing performance, size, and power requirements. If the macro implements a standard (for example, the IEEE 1394 interface), then its compliance must be stated, along with any exceptions or areas where the macro is not fully compliant to the published specification. VSIA suggests that, in addition to this information, the macro documentation include a section describing how the user can duplicate the development environment and verify the claims.

For soft IP, the deliverables include a reference library, complete scripts, and a verification environment, so these claims can be easily verified.

For hard IP, the end user does not have access to the GDSII, and so many of the claims are unverifiable. We recommend including actual measured values for timing performance and power in the user guide.

Exceptions to the coding/design guidelines

Any exceptions to the design and coding guidelines outlined in this manual must be noted in the user guide. It is especially important to explain any asynchronous circuits, combinational inputs, and combinational outputs.

Timing specifications and performance

Timing specifications include input setup and hold times for all input and I/O pins and clock-to-output delays for all output pins. Timing specifications for any combinational inputs/outputs must be clearly documented in the user guide. Timing for soft macros must be specified for a representative process.

CHAPTER 10

System Integration with Reusable Macros

This chapter discusses the process of integrating completed macros into the whole chip environment. The topics are:

- Integration overview
- Integrating soft macros
- Integrating hard macros
- Integrating RAMs and datapath generators
- Physical design

10.1 Integration Overview

The key issues in the final integration of the SoC design include:

- Logical design
- Synthesis and physical design
- Chip-level verification

This chapter addresses the first two subjects and the following chapter addresses verification issues.

10.2 Integrating Soft Macros

There are several key issues in designing a chip that uses soft macros provided by an external source:

- Selecting and/or specifying the macro
- Installing the macro into the design environment
- Designing and verifying the interfaces between the macro and the rest of the chip
- Functional verification of the macro in the chip
- Meeting timing requirements with the macro
- Meeting power requirements with the macro

10.2.1 Soft Macro Selection

The first step in selecting a macro from an external source, or in specifying a macro that is to be developed by an internal source, is to determine the exact requirements for the macro. For a standards-based macro, such as a PCI core or a IEEE1394 core, this means developing a sufficient understanding of the standard involved.

Once the requirements for the macro are fully understood, the choices can quickly be narrowed to those that (claim to) meet the functional, timing, area, and power requirements of the design. The most critical factors affecting the choice between several competing sources for a soft macro are:

Quality of the documentation

Good documentation is key to determining the appropriateness of a particular macro for a particular application. The basic functionality, interface definitions, timing, and how to configure and synthesize the macro should be clearly documented.

Robustness of the verification environment that accompanies the macro

Much of the value, and the development cost, of a macro lies in the verification suite. A rich set of models and monitors for generating stimulus to the macro and checking its behavior can make the overall chip verification much easier.

Robustness of the design

A robust, well-designed macro still requires some effort to integrate into a chip design. A poorly designed macro can create major problems and schedule delays. Verifying the robustness of a macro in advance of actually using it is difficult. A careful review of the verification environment and process is a first step. But for a macro to be considered robust, it must have been proven in silicon.

Ease of use

In addition to the above issues, ease of use includes the ease of interfacing the macro to the rest of the design, as well as the quality and user-friendliness of the installation and synthesis scripts.

10.2.2 Installation

The macro, its documentation, and its full design verification environment should be installed and integrated into your design environment much like an internally developed block. In particular, all components of the macro package should be under revision control. Even if you do not have to modify the design, putting the design under revision control helps ensure that it will be archived along with the rest of the design, so that the entire chip development environment can be recreated if necessary.

10.2.3 Designing with Soft Macros

Many soft macros are configurable through parameter settings. Designing with a soft macro begins with setting the parameters and generating the complete RTL for the desired configuration. Once this is done, you can instantiate the macro in your top-level RTL. The key issue here is to make sure that the interface between the macro and the rest of the design is correct. If the macro is well designed, the vast majority of problems with the macro will be in the interfaces to the rest of the chip. See the following chapter for more discussion on how to verify the macro in the system environment.

10.2.4 Meeting Timing Requirements

Robust and flexible synthesis scripts are the key to meeting timing requirements. Although the original designers may have synthesized the macro and verified that it met its timing specifications, they probably did so with a different configuration and a different target library. The scripts must be robust enough and flexible enough to synthesize the macro for the target configuration and library.

The integrator can greatly reduce the difficulty in meeting timing by selecting a gate array or standard cell technology that has a robust, synthesis-friendly library. Poorly designed or poorly characterized cells greatly hamper synthesis.

The integrator can also help by floorplanning the macro as a single unit. This makes it more likely that the wire loads in the actual design will be similar to the wire loads used by the macro designer in developing the macro. This may require that the macro appear high enough in the design hierarchy that the floorplanner can place it as a single entity.

10.3 Integrating Hard Macros

Designing a chip using hard macros provided by an external source involves several key issues:

- Selecting the macro
- Designing and verifying the interfaces between the macro and the rest of the chip
- Functional verification of the chip
- Timing verification of the chip
- Physical design issues

10.3.1 Hard Macro Selection

The first step in selecting a macro from an external source, or in specifying a macro that is to be developed by an internal source, is to determine the exact requirements for the macro. For microprocessor cores, this means developing an understanding of the instruction set, interfaces, and available peripherals.

Once the requirements for the macro are fully understood, the most critical factors affecting the choice between several competing sources for a hard macro are:

Quality of the documentation
Good documentation is key to determining the appropriateness of a particular macro for a particular application. The basic functionality, interface definitions, timing, and how to integrate and verify the macro should be clearly documented.

Completeness of the verification environment that accompanies the macro
In particular, functional, timing, synthesis, and floorplanning models must be provided.

Completeness of the design environment
If the macro is a microprocessor core, the vendor must supply or recommend a third-party supplier for the compilers and debuggers required to make the system design successful.

Robustness of the design
The design must have been proven in silicon.

Physical design limitations
Blockage and porosity of the macro—the degree to which the macro forces signal routing around rather than through the macro—must be considered. A design that uses many macros that completely block routing may result in very long wires between blocks, producing unacceptable delays.

10.3.2 Designing with Hard Macros

The first step in designing with a hard macro is to verify that the interfaces between the macro and the rest of the chip are correct. This involves simulating the rest of the chip using a cycle-accurate functional model for the macro.

10.3.3 Functional Verification

Functional verification of the entire chip can be done using either a cycle-accurate functional model or a more abstract model, such as a behavioral or ISA (Instruction Set Architecture) model for the macro. The advantage of the more abstract model is that simulation runs much faster, and more of the functional interactions of various blocks can be verified. This style of verification is appropriate only after the interfaces between the blocks have been robustly verified.

In some cases, a gate-level netlist may be available for a hard macro. This netlist will simulate much more slowly than the above models, but it will provide the most accurate timing model for the circuit. Most vendors do not currently provide gate-level netlists for hard macros for security reasons.

10.3.4 Timing Verification of the Chip

Timing verification for the whole chip requires a detailed, accurate timing model for the hard macro. The model must be sufficiently detailed that the interface timing between the macro and the other blocks in the design can be completely verified. A static timing analysis tool is typically used for this verification.

10.3.5 Physical Design Issues

Physical design issues that face the integrator include:

Placement and blockage
Typically, a hard macro limits or prevents the routing of signals through the macro itself. This means that placement of the macro can be critical in determining the routability and timing of the chip.

Clock distribution
Typically, the macro has its own internal clock tree. The delays in this tree must be compatible with the clock timing and skew of the rest of the chip.

Power and ground
The physical and electrical power and ground requirements of the macro must be met. This factor can also affect placement of the macro in the chip.

10.4 Integrating RAMS and Datapath Generators

Memories are a special case of the hard macro, and datapath generators are a variant of the soft macro. We discuss these two cases briefly in this section.

Large, on-chip memories are typically output from memory compilers. These compilers produce the functional and timing models along with the physical design information required to fabricate the memory. The issues affecting memory design are identical to those affecting hard macro designs, with the following additional issues:

*   The integrator typically has a wide choice of RAM configurations, such as single port or multi-port, fast or low-power, synchronous or asynchronous.

*   Asynchronous RAMs present a problem because generating a write clock often requires a gated clock. A fully synchronous RAM is preferred, if it meets the other design requirements.

*   Large RAMs with fixed aspect ratios can present significant blockage problems. Check with your RAM provider to see if the aspect ratio of the RAMs can be modified if necessary.

*   BIST is available for many RAM designs, and can greatly reduce test time and eliminates the need to bring RAM I/O to the chip's pins. But the integrator should be cautious because some BIST techniques do not test for data retention problems.

Datapath generators require a slightly different flow than standard soft macros, as described in the Module Compiler section of Chapter 6. To integrate a macro delivered as MC code, the integrator must:

*   Use the MC GUI or command line interface to configure the macro, define the synthesis constraints, and synthesize the gate-level netlist.

*   Feed the netlist into Design Compiler along with the other synthesizable designs for top-level synthesis.

*   Perform incremental synthesis as required to ensure that the interface timing between the macros meets the chip's requirements.

*   From here on, the flow for physical design is the same as for any soft macro.

*   The generated simulation model is cycle and bit accurate, and should be used for system-level verification, just as the functional model for a hard macro.

*   Timing and power analysis should be performed on the final netlist, just as for a soft macro.

10.5 Physical Design

Figure 10-1 outlines the process of integrating the various blocks into the final version of the chip and getting the chip through physical design.

This process starts with the top-level synthesis of the chip. At this point, all of the blocks should already have been synthesized and be known to meet timing. At the top level, synthesis should be required only to stitch together the top level netlist and refine the system-level interconnect: resizing buffers driving inter-block signals, fixing hold-time problems, and the like.

The inputs to the top-level synthesis include:

*   RTL (or a netlist synthesized from RTL) for the synthesizable blocks
*   Synthesis models for the hard macros and memories
*   Netlists for any modules generated from a datapath generator
*   Any libraries required, such as the DesignWare Foundation Library
*   Top level RTL that instantiates the blocks, the I/O pads, and top-level test structures

The synthesis models for the hard macros and memories include the timing and area information required to complete synthesis on the whole chip and to verify timing at the chip level.

The top-level test structures typically include any test controllers, such as a JTAG TAP controller or a custom controller for scan and on-chip BIST (Built-In Self Test) structures.

After the top-level netlist has been generated, scan cells should be inserted in the appropriate blocks for testability. An ATPG (Automatic Test Pattern Generator) tool can then be used to generate scan test vectors for the chip. Scan insertion is typically done by the synthesis tool or by a separate tool.

Similarly, if JTAG is required, JTAG structures should be added to the netlist. Typically, this step is also performed by a tool.

Once all the test structures are in place, a final timing analysis is performed to verify chip timing and, if necessary, an incremental synthesis is performed to achieve the chip timing requirements.

The final netlist, along with timing information, is then sent to a timing-driven floorplanning tool, along with floorplanning models for the hard macros and memories. A timing-driven floorplanner is essential for achieving fast timing convergence during the physical design of the chip.

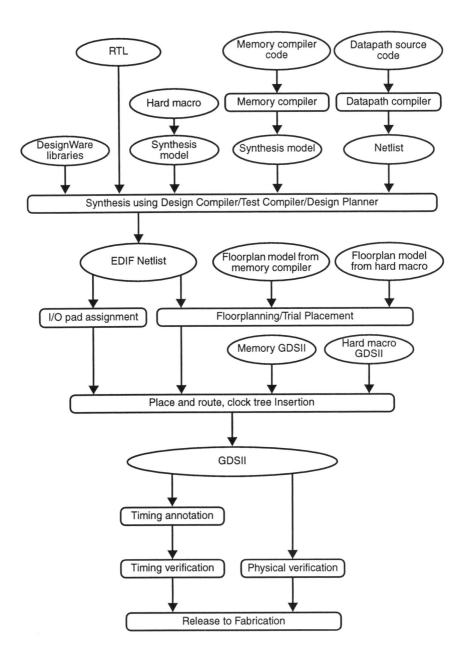

Figure 10-1 Integrating blocks into a chip

10.5.1 Achieving Timing Closure

An interactive loop between floorplanning and incremental synthesis allows the design team to achieve timing closure as rapidly as possible. The floorplanning tool performs a trial placement of the chip and generates estimated wirelengths and delays. These timing values are fed back to the synthesis tool and in-place optimization (IPO) is performed on paths that don't meet timing. The required netlist changes are fed into the floorplanning tool, which makes incremental changes to the placement.

Once timing goals are met, the final netlist and placement are fed into the place and route engine for final routing. Clock tree insertion is usually performed as part of the place and route process.

10.5.2 Verifying the Physical Design

Once the physical design is complete, a final timing verification is required. Timing information is extracted out of the final design database in the form of an SDF file and fed into a static timing analysis tool or simulator for final timing verification. If timing problems are found, the incremental synthesis-floorplanning-place and route process must be repeated.

Similarly, the physical designer performs a final physical design rule check on the design database. These checks include LVS (Layout vs. Schematic) and DRC (Design Rule Checking).

Once these steps are completed, the chip is ready for fabrication.

CHAPTER 11

System-Level
Verification Issues

This chapter discusses system-level verification, focusing on the issues and opportunities that arise when macros are integrated into a complete system on a chip. The topics are:

- The importance of verification
- Test plan
- Application-based verification
- Fast prototype testing
- Gate-level verification
- Verification tools
- Specialized hardware for system verification

11.1 The Importance of Verification

Verifying functionality and timing at the system level is probably the most difficult and important aspect of SoC design. It is the last opportunity to find conceptual, functional, and implementation errors before the design is committed to silicon. For many teams, verification takes 50 to 80 percent of the overall design effort.

For SoC design, verification must be an integral part of the design process from the start, along with synthesis, system software, bringup, and debug strategies. It cannot be an afterthought to the design process.

System verification begins during system specification. The system functional specification describes the basic test plan, including the criteria for completion (what tests must run before taping out). As the system-level behavioral model is developed, a testbench and test suite are developed to verify the model. Similarly, system software is developed and tested using the behavioral model rather than waiting for real hardware. As a result, a rich set of test suites and test software, including actual application code, should be available by the time the RTL and functional models for entire chip are assembled and the chip is ready for verification.

Successful (and rapid) system-level verification depends on the following factors:

- Quality of the test plan
- Quality and abstraction level of the models and testbenches used
- Quality and performance of the verification tools
- Robustness of the individual predesigned blocks

11.2 The Test Plan

The system-level verification strategy for a SoC design uses a divide-and-conquer approach based on the system hierarchy. This strategy consists of the following steps:

- Verify that the leaf nodes—the lowest-level individual blocks—of the design hierarchy are functionally correct as stand-alone units.
- Verify that the interfaces between blocks are functionally correct, first in terms of the transaction types and then in terms of data content.
- Run a set of increasingly complex applications on the full chip.
- Prototype the full chip and run a full set of application software for final verification.
- Decide when it is appropriate to release the chip to production.

11.2.1 Block Level Verification

Block level verification is described in detail in Chapter 7. It uses code coverage tools and a rigorous methodology to verify the macro at the RTL level as thoroughly as possible. A physical prototype is then built to provide silicon verification of functional correctness.

Any block in the SoC design that has not gone through this process, including silicon verification, is not considered fully verified as a stand-alone block. If the chip contains any such partially verified blocks, the first version of the chip must be considered a prototype. The first version of the chip on which a full set of application software is run is virtually assured of having bugs that require a redesign of the chip before release to production.

11.2.2 Interface Verification

Once the individual blocks have been verified as stand-alone units, the interfaces between blocks need to be verified. These interfaces usually have a regular structure, with address and data buses connecting the blocks and some form of control—perhaps a request/grant protocol or a request/busy protocol. The connections between blocks can be either point-to-point or on-chip buses.

Because of the regular structure of these interfaces, it is usually possible to talk about *transactions* between blocks. The idea is that there are only a few permitted sequences of control and data signals; these sequences are called transactions and only the data (and data-like fields, such as address) change from transaction to transaction.

Transaction Verification

Interface testing begins by listing all of the transaction types that can occur at each interface, and systematically testing each one. If the system design restricts transactions to a relatively small set of types, it is fairly easy to generate all possible transaction types and sequences of transaction types and to verify the correct operation of the interfaces to these transactions. Once this is done, all that remains is to test the behavior of the blocks to different data values in the transactions. Thus, a simple, regular communication architecture between blocks can greatly reduce the system verification effort.

In the past, this transaction checking has been done very informally by instantiating all the blocks in the top level RTL, and then using a testbench to create activity within the blocks and thus transactions between blocks. If the overall behavior of the system is correct, perhaps as observed at the chip's primary I/O or in the memory contents, then the chip—and thus the interfaces—were considered to be working correctly.

There are several changes that can be made to improve the rigor of transaction checking. First of all, as shown in Figure 11-1(b), you can add a monitor to check the transactions directly. This monitor can be coded behaviorally and be used during simulation. However, if it is designed at the RTL level, it can be used for cycle-based simulation and emulation. It can also be built into the actual chip to facilitate bring-up and debug, as well as runtime error checking. For a chip such as that shown in Figure 11-1(a), with point-to-point interconnections, it is possible to build some simple transaction checking into the interface module of each block.

This monitor approach improves observability during transaction testing, but it is also possible to improve controllability. If we use a simple, transaction-generating bus functional model instead of a fully functional model for the system blocks, we can generate precisely the transactions we wish to test, in precisely the order we want. This approach can greatly reduce the difficulty of developing transaction verification tests and can reduce simulation runtime as well.

One way to accomplish this in practice is shown in Figure 11-1. Here we have separated the interface portions of each block from the internals. We then combine the behavioral model for the internals of each block with the actual RTL description of the interface portion of the block.

Data or Behavioral Verification

Once the transactions have been verified, it is necessary to verify that the block behaves correctly for all values of data and all sequences of data that it will receive in actual operation. In most chips, generating the complete set of these data is impossible because of the difficultly in controlling the data received by any one block.

The approach shown in Figure 11-1 helps here as well. It uses simple, behavioral models, essentially bus functional models (BFMs), for all blocks except the block under test. You use the full RTL design for the block under test and generate the desired data sequences and transaction from the BFMs. You can construct test cases either from your knowledge of the system or by random generation.

Automatic checking of the block's behavior under these sequences of transactions is nontrivial and depends on how easy it is to characterize the correct behavior of the block. For complex blocks, the only automated checking that is practical may be verifying that none of the state machines hang, and that no illegal transactions are generated as a result of legal inputs.

This test method often reveals that the block responds correctly to data sequences that the designer expected the block to receive, but that there are some (legal or illegal) sequences that can occur in the actual system to which the block does not respond correctly.

Another method for dealing with the problem of unanticipated or illegal inputs is to design a checker into the block interface itself. This checker can suppress inputs that are not legal and prevent the block from getting into incorrect states. This approach has been used effectively in high-reliability system designs.

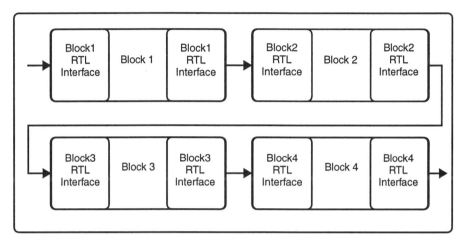

(a) Chip with point-to-point interfaces

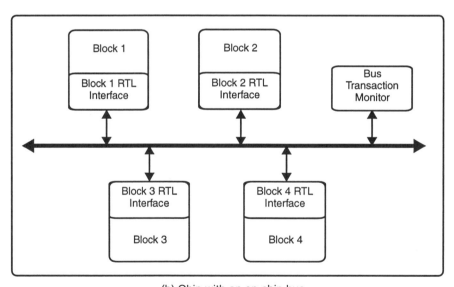

(b) Chip with an on-chip bus

Figure 11-1 System verification using interface testing

11.3 Application-Based Verification

Once the basic functionality of the system has been verified by the transaction testing described in the previous section, system verification consists of exercising the entire design, using a full functional model for most, if not all, of the blocks. The ultimate goal of this aspect of verification is to try to test the system as it will actually be used. That is, we come as close as we can to running actual applications on the system.

Verification based on running real application code is essential for achieving a high quality design. However, this form of verification presents some major challenges. Conventional simulation, even at the RTL level, is simply not fast enough to execute the millions of vectors required to run even the smallest fragments of application code, much less to boot an operating system or test a cellular phone.

There are two basic approaches to addressing this problem:

- Increase the level of abstraction so that software simulators running on work-stations run faster.
- Use specialized hardware for performing verification, such as emulation or rapid-prototyping systems.

This section addresses the first approach: how to use abstraction and other mechanisms to speed conventional simulation techniques. Subsequent sections address the second approach.

11.3.1 A Canonical SoC Design

The types of abstraction techniques we can use depend on the nature of the design, so it is useful to use a specific design as an example. Fortunately, most large chips are converging to an architecture that looks something like the chip design shown in Figure 11-2, the canonical SoC design described in Chapter 2.

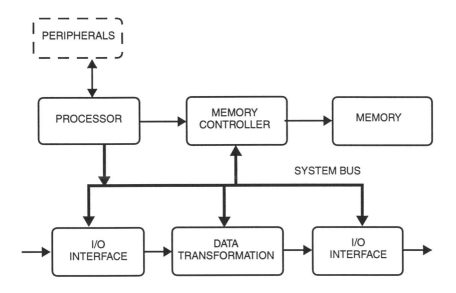

Figure 11-2 Canonical SoC Design

11.3.2 A Testbench for the Canonical Design

Figure 11-3 shows a possible testbench environment for verifying the canonical design. The key features of this verification environment are:

- The full RTL model is used as the simulation model for most of the functional blocks.
- Behavioral or ISA (Instruction Set Architecture) models may be used for memory and the microprocessor.
- Bus functional models and monitors are used to generate and check transactions with the communication blocks.
- It is possible to generate real application code for the processor and run it on the simulation model.

With this test environment, we can run a set of increasingly complex application tests on the system. Initially, full functional models for the RAM and microprocessor are used to run some basic tests to prove that the system performs the most basic functions. The slow simulation speeds of this arrangement mean that we can do little more than check that the system is alive and find the most basic system bugs. Errors are detected manually (by looking at waveform displays), by means of the bus monitor, and by the sequence monitor on the communication port. At this level of abstraction, we are probably simulating at a rate of tens of system clocks per second.

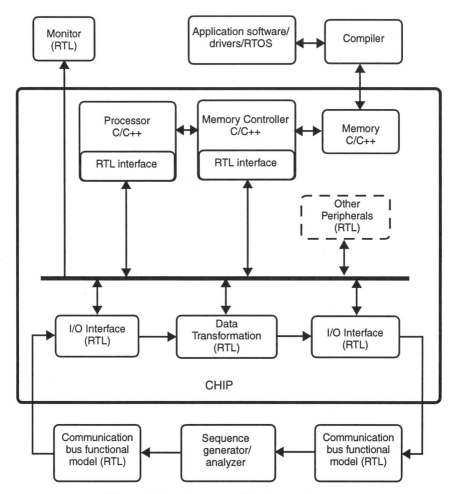

Figure 11-3   System verification environment

Behavioral models are now substituted for the memory and microprocessor. These models can be high-level C/C++ models that accurately model the instruction set of the processor, but abstract out all implementation detail. These models are often called ISA (Instruction Set Architecture) models. Another approach is to code a very high-level, behavioral model in Verilog or VHDL, abstracting out much of the cycle-by-cycle details, but retaining the basic functionality of the processor. If enough timing detail is retained so that the bus transactions at the I/O port of the process are accurate on a clock-cycle by clock-cycle basis, the model is often referred to as a cycle-accurate model.

Using these behavioral models for the RAM and processor, real code is compiled and loaded into the RAM model and the processor model executes this code. At the same time, representative data transactions are generated at the communication interfaces of the chip, usually by bus functional models. For instance, if the data transformation block is an MPEG core, then we can feed in a digital video stream.

Using C/C++ models for both the processor and memory dramatically improves simulation speed over full RTL simulation. In designs like our canonical example, most cycles are spent entirely in the processor, executing instructions, or in accessing memory. With these abstractions, execution speeds in the thousands of device cycles per second can be achieved. Operating on top of this environment, hardware/software cosimulation packages allow the engineer to run a software debugger, the ISA software, and an RTL simulator simultaneously.

Most system-level hardware and software bugs can be detected and fixed at this stage. To complete software debug, it may be necessary to develop an even more abstract set of models to improve simulation speed further. In our example, we could substitute a C++ model for the RTL model of the data transformation block, and achieve very high simulation speeds.

To complete hardware debug, however, we need to lower our abstraction level back to RTL. The lack of detail in our ISA/behavioral models undoubtedly masks some bugs. At this point, we can run some real code on the RTL system model and perhaps some random code as well for testing unforeseen sequences. But simulation speed prohibits significant amount of real application code from being run at the RTL level.

During this debug phase, as we run application code on a high-level model and targeted tests on the RTL model, the bug rate follows a predictable curve. Typically, the bug rate increases during the first part of this testing, reaches a peak, and then starts declining. At some point on the downward slope, simulation-based testing is providing diminished returns, and an alternate method must be found to detect the remaining bugs.

11.4 Fast Prototype vs. 100 Percent Testing

For most design teams, a key goal is to have first silicon be fully functional. This goal has motivated the functional verification plan and simulation strategies. To date, most teams have been fairly successful. According to some estimates, about 90% of ASIC designs work right the first time, although only about 50% work right the first time in the system. This higher failure rate probably results from the fact that most ASIC design teams do not do system-level simulation.

With the gate count and complexity of SoC designs, it is not clear that the industry can maintain this success rate. Assume that, in a 100k gate design with today's verification technology, there is a 10% chance of a serious bug. Then for a 1M gate design, consisting of ten such modules comparably verified, the probability of no serious bugs is:

$$P_{bug\text{-}free} = .9^{10} = .35$$

Design reuse can also play an important role. If we assume that a 1M gate design consists of ten 100k blocks, with two designed from scratch (90% chance of being bug-free) and eight reused (say, 98% chance of being bug-free), then for the overall chip:

$$P_{bug\text{-}free} = .9^2 * .98^{\,8} = .69$$

But to achieve a 90% probability of first-silicon success, we need to combine design reuse with a verification methodology that will either get individual blocks to a 99% or allow us to verify the entire chip to the 90% level.

Running significant amounts of real application code is the only way to reach this level of confidence in an SoC design. For most designs, this level of testing requires running at or near real time speeds. The only available technologies for achieving this kind of performance involve some form of rapid prototyping.

The available options for rapid prototyping include:

- FPGA or LPGA prototyping
- Emulation-based testing
- Real silicon prototyping

11.4.1   FPGA and LPGA Prototyping

For small designs, it is practical to build an FPGA or Laser Programmable Gate Array (LPGA, such as the one provided by Chip Express) prototype of the chip. FPGAs have the advantage of being reprogrammable, allowing rapid turnaround of bug fixes. LPGA prototypes can achieve higher gate counts and faster clock speeds, but are expensive to turn. Multiple iterations of an LPGA design can be very costly, but can be done quickly, usually within a day or two.

Both FPGAs and LPGAs lag state-of-the-art ASIC technologies in gate count and clock speed by significant amounts. They are much more appropriate for prototyping individual blocks or macros than for prototyping SoC designs.

A number of engineering teams have used multiple FPGAs to build a prototype of a single large chip. This approach has at least one major problem: the interconnect is difficult to design and almost impossible to modify quickly when a bug fix requires repartitioning of the design between devices.

Rapid prototyping systems from Aptix address this problem by using custom, programmable routing chips to connect the FPGAs. This routing can be performed under software control, providing a very flexible fast prototyping system.

11.4.2   Emulation Based Testing

Emulation technology such as that provided by Mentor Graphics and QuickTurn grew out of attempts to provide a better alternative to using a collection of FPGAs for rapid prototyping of large chips. They provide programmable interconnect, fixed board designs, relatively large gate counts, and special memory and processor support. Recent developments in moving from FPGAs to processor-based architectures have helped to resolve partitioning and interconnect problems.

Emulation can provide excellent performance for large-chip verification if the entire design can be placed in the emulation engine itself. If any significant part of the circuit or testbench is located on the host, there is significant degradation of performance.

For our canonical design, we need to provide emulation-friendly models for the RAM, microprocessor, BFMs, monitor, and sequence generator/checker. Developing these models late in the design process can be so time consuming as to negate the benefit of emulation. It is much better to consider the requirements of emulation from the beginning of the project and to work with the memory and hard macro providers to provide these models. Similarly, the requirements of emulation must be considered in the design of the BFMs and monitors.

If executed correctly, emulation can provide simulation performance of one to two orders of magnitude less than real time, and many orders of magnitude faster than simulation.

11.4.3   Silicon Prototyping

If an SoC design is too large for FPGA/LPGA prototyping and emulation is not practical, then building a real silicon prototype may be the best option. Instead of extending the verification phase, it may be faster and easier to build an actual chip and debug it in the system.

To some extent this approach is just acknowledging the fact that any chip fabricated without running significant amounts of real code must be considered a prototype.

That is, there is a high probability that engineering changes will be required before release to production.

The critical issue in silicon prototyping is deciding when one should build the prototype. The following is a reasonable set of criteria:

- The bug rate from simulation testing should have peaked and be on its way down.
- The time to determine that a bug exists should be much greater than the time to fix it.
- The cost of fabricating and testing the chip is on the same order as the cost of finding the next n bugs, where n is the anticipated number of critical bugs remaining.
- Enough functionality has been verified that the likely bugs in the prototype should not be severe enough to prevent extensive testing of other features. The scenario we want to avoid is building a prototype only to find a critical bug that prevents any useful debug of the prototype.

There are a number of design features that can help facilitate debug of this initial prototype:

- Good debug structures for controlling and observing the system, especially system buses
- The ability to selectively reset individual blocks in the design
- The ability to selectively disable various blocks to prevent bugs in these blocks from affecting operation of the rest of the system

11.5 Gate-Level Verification

The final gate-level netlist must be verified for both correct functionality and for timing. A variety of techniques and tools can be used for this task.

11.5.1 Sign-Off Simulation

In the past, gate-level simulation has been the final step before signing off an ASIC design. ASIC vendors have required gate-level simulation and parallel test vectors as part of signoff, using the parallel vectors as part of manufacturing test. They have done this even if a full scan methodology was employed.

Today, for 100k gate and larger designs, signoff simulation is typically done running Verilog simulation with back-annotated delays on hardware accelerators from IKOS. Running full-timing, gate-level simulations in software simulators is simply not feasible at these gate counts. Even with hardware accelerators, speeds are rarely faster than a few hundred device cycles per second.

RTL sign-off, where no gate-level simulation is performed, is becoming increasingly common. However, most ASIC vendors still require that all manufacture-test vectors submitted with a design be simulated on a sign-off quality simulator with fully back-annotated delay information and all hazard checking enabled. Further, they require that these simulations be repeated under best case, nominal case, and worst case conditions. This has the potential to be a resource intensive task.

This requirement is rapidly becoming problematic for the following reasons:

- Thorough, full timing simulation of a million-gate ASIC is not possible without very expensive hardware accelerators, and even then it is very slow.
- Parallel vectors typically have very low fault coverage (on the order of 60 percent) unless a large and expensive effort is made to extend them. As a result, they can be used only to verify the gross functionality of the chip.
- Parallel vectors do not exercise all the critical timing paths, for the same reason they don't achieve high fault coverage. As a result, they do not provide a sufficient verification that the chip meets timing.

As a result of these issues, the industry is moving to a different paradigm. The underlying problems traditionally addressed by gate-level simulation are:

- Verification that synthesis has generated a correct netlist, and that subsequent operations such as scan and clock insertion have not changed circuit functionality
- Verification that the chip, when fabricated, will meet timing
- A manufacturing test that assures that a chip that passes test is free of manufacturing defects

These problems are now too large for a single solution, such as gate-level simulation. Instead, the current methodology uses separate approaches to address each issue:

- Formal verification is used to verify correspondence between the RTL and final netlist.
- Static timing analysis is used to verify timing.
- Some gate-level simulation, either unit-delay or full timing, is used to complement formal verification and static timing analysis.
- Full scan plus BIST provides a complete manufacturing test for functionality. Special test structures, provided by the silicon vendor, are used to verify that the fabricated chip meets timing and other analog specifications.

11.5.2 Formal Verification

Formal verification uses mathematical techniques to prove the equivalence of two representations of the circuit. Typically, it is used to compare the gate-level netlist to the RTL for a design. Because it uses a static, mathematical method of comparison, formal verification requires no functional vectors. Thus, it can compare two circuits much more quickly than can be done with simulation, and with much greater accuracy. Formal verification is available from a variety of vendors; one such tool is Synopsys Formality.

Formality compares two design by reading them into memory and then applying formal mathematical algorithms on their data structures. The designs can be successfully compared as long as they have the same synchronous functionality and correlating state holding devices (registers or latches). The two circuits are considered equivalent if the functionality is the same at all output pins and at each register and latch.

Formal verification can be used to check equivalence between the original RTL and:

- The synthesized netlist
- The netlist after test logic is inserted. For scan, this is quite straightforward; for on-chip JTAG structure, some setup is required, but the equivalence can still be formally verified.
- The netlist after clock tree insertion and layout. This requires comparing the hierarchical RTL to the flattened netlist.
- Hand edits. Occasionally engineers will make a last-minute hand edit to the netlist to modify performance, testability, or function.

One key benefit of formal verification is that it allows the RTL to remain the golden reference for the design, regardless of modifications made to the final netlist. Even if the functionality of the circuit is changed by a last minute by editing the netlist, the same modification can be retrofitted into the RTL and the equivalence of the modified RTL and netlist can be verified.

11.5.3 Gate-Level Simulation with Unit-Delay Timing

Unit-delay simulation involves performing gate-level simulation with unit delay for each gate. It is much faster than full-timing simulation, but much slower than RTL simulation.

Unit-delay simulations can be used to verify that:

- The chip initializes properly (reset verification).
- The gate implementation functionally matches the RTL description (functional correctness).

Gate-level simulation complements formal verification. Dynamic simulations are rarely an exhaustive test of equivalence, but simulation is necessary to validate that an implementation's behavior is consistent with the simulated behavior of the RTL source.

Because it can be time-consuming and resource-intensive, it is usually good to begin unit-delay simulation as soon as you complete a netlist for your chip, even though the chip may not meet timing.

11.5.4 Gate-Level Simulation with Full Timing

Full-timing simulation on large chips is very slow, and should be used only where absolutely necessary. This technique is particularly useful for validating asynchronous logic, embedded asynchronous RAM interfaces, and single-cycle timing exceptions. In a synchronous design, these problem areas should not exist, or should be isolated so they are easily tested.

These tests should be run with the back-annotated timing information from the place and route tools, and run with hazards enabled. They should be run with worst case timing to check for long paths, and with best-case timing to check for minimum path delay problems.

11.6 Choosing Simulation Tools

Determining what simulation tools to use at the various levels of verification is a decision that must be made early in the design process. Effective use of the various available verification technologies can help minimize verification time while improving the quality of the verification itself. Using different tools at different stages of the design does present the problem that the blocks and the various testbenches must be designed to work with all the verification tools with which they will be used.

This decision affects testbench design at all levels of verification. Figure 11-4 shows one reasonable selection of tools, as described below:

- For the lowest level of subblock verification, an *interpreted, event-based simulator* is the most useful. At this point in the design process, fast compile time and good debug support are more important than fast runtime. Testbenches that are not going to be used at higher levels of integration can be behavioral; typically these are faster to write and to modify.

- For block integration, the design team will probably want to migrate to a *compiled simulator* as soon as the code is relatively stable. At this point, bugs are more subtle and require longer runtimes to find. A compiled, event-driven simulator provides faster run times, but slower compiles, than an interpreted simulator. A *compiled, cycle-based simulator* can provide even faster run times, but with some restrictions on coding style. Cycle-based simulators support both behavioral and synthesizable testbenches, but behavioral testbenches run more slowly than synthesizable testbenches. Testbenches that will migrate to emulation with the design should be written as synthesizable testbenches.

- When cycle-based simulation becomes too slow to find additional bugs, the design needs to be moved to an *emulator*. Initially, cosimulation is used to allow the testbench and most of the blocks to remain in the cycle-based simulator. Blocks are migrated to the emulator one at a time, as they become stable and relatively bug-free. The speed advantage of the emulator becomes greater as more blocks are moved to it, but the speed-up over cycle-based simulation is limited as long as any code remains outside the emulator.

At some point, the design is stable enough that it is necessary to run significant amounts of real code on the design. The only way to achieve this is to move the entire design, including testbench, to the emulator. This requires a synthesizable testbench; if this requirement has been considered from the start, this should be a reasonably straightforward conversion.

11.7 Specialized Hardware for System Verification

Design teams have long recognized the limitations of software simulators running on workstations. Simulation has never provided enough verification bandwidth to do really robust system simulation. Over the last fifteen years there have been numerous efforts to address the needs of system simulation through specialized hardware systems for verification.

Early efforts focused on hardware accelerators. Zycad introduced the first widely-available commercial accelerators in the early 1980's; in the early 1990's Ikos introduces competitive systems based on somewhat similar architectures. The Zycad systems provided very fast fault simulation; at the time fault simulation of large chips was not really practical with software simulators. These systems were also used for gate-level system simulation. Ikos systems focus exclusively on system-level simulation.

These accelerators map the standard, event-driven software simulation algorithm onto specialized hardware. The software data structures used to represent information about gates, netlists, and delays are mapped directly into high-speed memories. The algorithm itself is executed by a dedicated processor that has the simulation algorithm hardwired into it. A typical system consists of anywhere from 4 to over a hundred of these processors and their associated memory. These systems are faster than workstations because each processor can access all the needed data structures at the same time and operate on them simultaneously. Additional performance results from the parallel execution on multiple processors.

The introduction of FPGAs in the 1980's made possible another kind of verification system: emulation. These systems partition the gate-level netlist into small chunks and map them onto FPGAs; they use additional FPGAs to provide interconnect routing. These systems can execute many orders of magnitude faster than hardware accelerators. Large circuits that run tens of cycles per second on software simulators might run hundreds or a few thousand of cycles per second on a hardware accelerator. These same circuits run at hundreds of thousands of cycles per second on emulation systems.

Different simulation tools are recommended for each stage

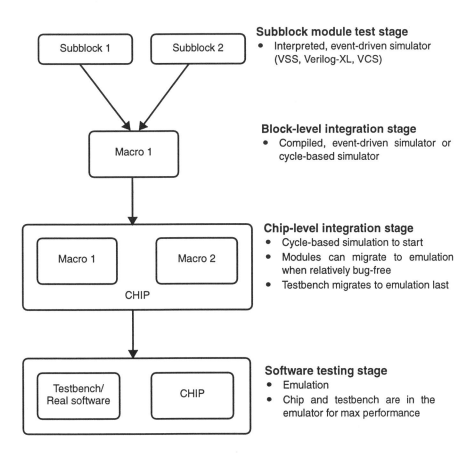

Figure 11-4   Recommended simulation tools for different design stages

Emulation systems achieve their high performance because they are essentially building a hardware prototype of the circuit in FPGAs.

Emulation systems, however, have a number of shortcomings:

- They operate on gate-level netlists. Synthesis is typically used to generate this netlist from the RTL. Any part of the circuit that is coded in non-synthesizable code, especially testbenches, must run on the host workstation. This considerably slows emulation: a circuit with a substantial part executed on the workstation may run as much as two orders of magnitude slower than one with the entire circuit in the emulator.
- The partitioning of the circuit among numerous FPGAs, and dealing with the associated routing problems, presents a real problem. Poor utilization and routing inefficiencies result in the need for very large numbers of FPGAs to emulate a reasonable sized chip. The resulting large systems are very expensive and have so many mechanical components (chips, boards, and cables) that they tend to experience reliability problems.
- The use of FPGAs tends to make controlling and observing individual nodes in the circuit difficult. Typically, the circuit has to be (at least partially) recompiled to allow additional nodes to be traces. This makes debugging in the emulation environment difficult.

The first problem remains an issue today, but important progress has been made on the second and third problems. New systems, such as those from Mentor Graphics (the Accelerated Verification System) and from QuickTurn have moved from a pure FPGA-based system to a custom chip/processor-based architecture. Where previous systems had arrays of FPGAs performing emulation, the new systems have arrays of special purpose processors. These processor-based systems usually use some form of time-slicing: the processor emulates some gates on one cycle, additional gates on the next cycle. Also, the interconnect between processors is time-sliced, so that a single physical wire can act as several virtual wires. This processor-based approach significantly improves the routing and density problems seen in earlier emulation systems.

The processors used on these new systems also tend to have special capabilities for storing stimulus as well as traces of nodes during emulation. This capability helps make debug in the emulation environment much easier.

These new systems hold much promise for addressing the problems of high-speed system verification. The success of these systems will depend on the capabilities of the software that goes with them: compilers, debuggers, and hardware/software cosimulation support. These systems will continue to compete against much less expensive approaches: simulation using higher levels of abstraction and rapid prototyping.

The rest of this chapter discusses emulation in more detail, using the Accelerated Verification System from Mentor as an example.

11.7.1 Accelerated Verification Overview

Figure 11-5 shows the major components of Mentor Graphics' Accelerated Verification System process. The components are:

Models
RTL blocks and soft IP are synthesized and mapped onto the emulation system hardware. Memory blocks are compiled and emulated on dedicated memory emulation hardware.

Physical environment
Hard macros (IP cores) that have bonded-out chips, can be mounted on special board and interfaced directly to the rest of the emulation system. Similarly, hardware testbenches, such as signal generators, can be connected directly to the emulation system.

In-circuit verification
The emulation system can be interfaced directly to a board or system to provide in-circuit emulation. Thus, an application board can be developed and debugged using an emulation model of the chip.

System environment
A software debug environment (XRay debugger) and a hardware/software co-simulation environment (Seamless CVE) provide the software support necessary for running and debugging real system software on the design.

Testbenches
Behavioral RTL testbenches can be run on the host and communicate with the emulation system. Note that running any significant amount of code on the host will slow down emulation considerably.

Stimulus
Synthesizable testbenches can be mapped directly onto the emulation system and run at full emulation speeds. Test vectors can be stored on special memories in the emulation system, so that they too can be run at full speed.

These components combine to provide all the capabilities that designers need to verify large SoC designs including:

- RTL acceleration
- Software-driven verification at all levels in the design cycle
- In-circuit verification to ensure the design works in context of the system
- Intellectual property support

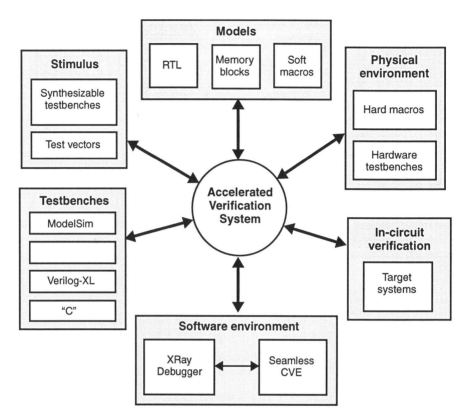

Figure 11-5   Mentor Graphics Accelerated Verification System process

11.7.2   RTL Acceleration

Designers continue to use software simulators like ModelSim, VSS, or Verilog-XL to debug their designs, but a threshold is reached where simulator performance becomes a major bottleneck for functional verification at the RTL level, especially for large SoC designs. This threshold will vary based on the design team and the verification environment. As the RTL functional simulations reach duration of more than 6-8 hours, it will become more efficient to compile the design and run it in an emulator. As an example, an RTL design that may only take several minutes to compile on a simulator, but runs for eight hours, may compile in 30 minutes on the emulator and run in a matter of seconds.

Thus, at some point in the design cycle, system simulation and debug may be more appropriately done on the emulator than on the simulator. For this debug strategy to be effective, however, we need an RTL symbolic debug environment that provides users with complete design visibility, real time variable access, and support for enumerated types and state variable assignments.

11.7.3 Software Driven Verification

As the software content in SoC designs increases and design cycles shrink, hardware/software co-development and software-driven verification become increasingly important. Software-driven verification plays two key roles in an SoC verification strategy:

- Verification of the hardware using real software
- Verification of the software using (an emulation model of) real hardware, well before the actual chip is built

Traditionally, using the software to verify the hardware has been confined to very late in the design cycle using breadboards, or early prototype runs of silicon. With reusable hardware and software blocks, it is possible to assemble an initial version of the RTL and system software very quickly. With the new emulation systems, it is possible to test this software on an emulation model of the hardware running at near-real-time speeds. Incremental improvements to both the hardware and software can then be made and tested, robustly and quickly.

In particular, the high performance of emulation systems allows the design team to:

- Develop and debug the low level hardware device drivers on the virtual prototype with hardware execution speeds that can approach near real-time
- Boot the operating system, initialize the printer driver, or place a phone call at the RTL phase of the design cycle

11.7.4 Traditional In-Circuit Verification

As the design team reaches the end of the design cycle, the last few bugs tend to be the most challenging to find. In-circuit testing of the design can be key tool at this stage of verification because the ultimate verification testbench is the real working system. One large manufacturer of routers routinely runs its next design for weeks on its actual network, allowing the router to deal with real traffic, with all of the asynchronous events that are impossible to model accurately.

In the case of systems that operate with the real asynchronous world, with random events that might occur only on a daily or weekly basis, complete information capture is essential. The emulation system provides a built-in logic analyzer that records every signal in the design during every emulation run, using a powerful triggering mechanism. This debug environment allows designers to identify and correct problems without having to repeat multiple emulation runs.

11.7.5 Support for Intellectual Property

Mentor Graphics Accelerated Verification System offers very secure encryption mechanisms and advanced macro compile capabilities that allow IP developers to have complete control over what parts of the IP modules are visible to the end user.

11.7.6 Design Guidelines for Accelerated Verification

Most of the guidelines for Accelerated Verification are identical to guidelines for design reuse listed in Chapter 5. These include:

Guideline – Use a simple clocking scheme, with as few clock domains as possible. Emulation works best with a fully synchronous, single-clock design.

Guideline – Use registers (flip-flops), not latches.

Guideline – Do not use combinational feedback loops, such as a set-reset latch in cross-coupled gates.

Guideline – Do not instantiate gates, pass transistors, delay lines, pulse generators, or any element that depends on absolute timing.

Guideline – Avoid multi-cycle paths.

Guideline – Avoid asynchronous memory. Use the modeling techniques described in Chapter 5 to model the asynchronous memory as a synchronous memory.

The following guidelines are requirements specific to emulation:

Guideline – Hierarchical, modular designs are generally easier to map to the emulation hardware than flat designs. The modularity helps reduce routing between processors.

Guideline – Large register arrays should be modeled as memories, to take advantage of the special memory modeling hardware in the emulation system.

CHAPTER 12

Data and Project Management

This chapter discusses tools and methodologies for managing the design database for macro design and for system design. The topics are:

- Data management
- Project management

12.1 Data Management

Data management issues include revision control, bug tracking, regression testing, managing multiple sites, and archiving the design project.

12.1.1 Revision Control Systems

A strong revision control system is essential for any design project. A good revision control system allows the design team to:

- Keep all source code (including scripts and documentation) in one centralized repository that is regularly backed up and archived
- Keep all previous versions of each file
- Identify quickly changes between different revisions of files
- Take a snapshot of the current state of the design and label it

RCS, SCCS, and Clearcase are examples of tools with which revision control systems can be built. Additional scripts and processes are typically used to create complete revision control system.

The most common paradigm is for each designer to be able to check out the entire design structure and recreate it locally, either by copying files or creating pointers to them. The designer then works and tests locally before checking the design files back into the central repository.

There are two different models for controlling this check-in process: the always-broken and always-working models.

The Always-Broken Model

In the *always-broken* model, each designer works and tests locally and then all the designers check in their work at the same time. The team then runs regression tests on the whole design, fixing bugs as they appear.

There are two problems with this model. First, when regressions tests fail, it is not clear whose code broke the design. If there are complex inter-dependencies between the modules, debugging regression failures can be difficult and time consuming.

The second problem with this model is that there tends to be a long integration period during which the design is essentially broken. No new design work can be done during this integration and debug phase because designers cannot check out a known-working copy of latest version of the design.

The Always-Working Model

The *always-working* model overcomes the major problems presented by the always-broken model. For the initial integration of the design, when separate modules are first tested together, the always-working model is the same as the always-broken model. Everyone checks in the initial version of the blocks and a significant debug effort ensues. In some designs, it may be possible to integrate a subset of the whole design, and then add additional blocks once the first subset is working. This approach greatly reduces the debug effort.

Once an initial baseline for the design is established, the always-working model uses the following check-in discipline:

- Only one designer can have a given block checked out for editing.
- When a block is being checked in, the entire central repository is locked, blocking other designers from checking modules in.

- The designer then runs a full regression test with the existing design plus the modified block.
- Once the regression tests pass, the designer checks in the block and removes the lock.

This model ensures that the entire design in the central repository always passes regression testing; that is, it is always working. It also greatly reduces the debug effort because only one new module at a time is tested.

We recommend the always-working model of revision control.

12.1.2 Bug Tracking

An effective bug tracking system is essential for rapid design of complex blocks and systems. A central database that collects all known bugs and desired enhancements lets the whole team know the state of the design and prevents designers from debugging known problems multiple times. It also assures that known problems are not forgotten, and that any design that is shipped with known bugs can include documentation for the bugs.

Another key use for bug tracking is bug rate tracking. In most projects, the bug rate follows a well-defined curve, reaching a peak value early in the integration phase and then declining as testing becomes more robust. The current bug rate and the position on this curve help define the most effective testing and debug strategy for any phase of the project, and help determine when the chip is ready to tape out.

Formal bug tracking usually begins when integration begins; that is, as soon as the work of two or more designers is combined into a larger block. For a single engineer working on a single design, informal bug tracking is usually more effective. However, some form of bug tracking is required at all stages of design.

12.1.3 Regression Testing

Automated regression testing provides a mechanism for quickly verifying whether changes to the design have broken a previously-working feature. A good regression testing system automates the addition of new tests, report generation, and distribution of simulation over multiple workstations. It should also highlight differences in output files between passing and failing tests, to help direct debug efforts.

12.1.4   Managing Multiple Sites

Many large projects involve design teams located at multiple sites, sometimes scattered across the globe. Effective data management across these sites can facilitate cooperation between the teams.

Multiple site data management starts with a high-speed link between sites; this is essential for sharing data. The revision control central repository must be available to all sites, as well as bug tracking information. Regression test reports must be available to all sites.

The key to managing a multi-site project is effective communication between the individual engineers working on the project. Email, voicemail, and telephones have been the traditional tools for this. Technology is now readily available for desk-top video conferencing with shared displays and virtual whiteboards. All of these techniques are needed to provide close links between team members.

One management technique that helps make these technology-based solutions more effective is to get the entire team in one place for an initial planning and teambuilding session. Once team members get to know the people at other sites, the daily electronic communication becomes much more effective.

12.1.5   Archiving

At the end of any design project, it is necessary to archive the entire design database, so that it can be re-created in the future, either for bug fixes or for enhancements to the product. All aspects of the design must be archived in one central place: documentation, source code, all scripts, testbenches, and test suites. All the tools used in the design must also be archived in the revision control system used for the design. If these tools are used for multiple projects, obviously one copy is enough, and the tool archives can be kept separate from the design archive.

The above observation may seem obvious, but let me interject a personal note. Several years ago, I was hired to do the next generation design of an existing large system. The first thing I did was to try to find the people who had worked on the previous generation design and to find the design archive.

Well, the designers had all moved on to other companies. The system architect was still with the company but busy with another project, and he wasn't all that familiar with the detailed design anyway. The design files were scattered across a wide variety of machines, some of which were obsolete machines cobbled together from spare parts and whose disks were not backed up! Worse than that, I found several copies of the design tree, with inconsistent data. It was impossible to collect even a majority of design files and know with confidence that they were the latest versions.

In the middle of the previous design effort, the team had changed silicon vendors and tools. The HDL for the design was *almost* accurate, but some (unspecified) changes were made directly to the netlist. This scenario is the manager's nightmare. It took months to recover the design archive to the point where the new design effort could begin; in fact, close to half the design effort consisted of recreating the knowledge and data that should have been archived at the end of the design.

12.2 Project Management

There are many excellent books on project management, and we will not attempt to cover the subject in any depth. However, there are a several issues that are worth addressing.

12.2.1 Development Process

An ISO 9000-like development process, where processes are documented and repeatable, can help considerably in producing consistently high-quality, reusable designs. Such a process should specify:

- The *product development lifecycle*, outlining the specific phases of the design process and the criteria for transitioning from one phase to another
- What *design reviews* are required at the different stages of the design, and how the design reviews will be conducted
- What the *sign-off process* is to complete the design
- What *metrics* are to be tracked to determine the completeness and robustness

Two key documents are used to communicate with the rest of the community during the course of macro design. These documents are the *project plan* and the *functional specification*. These are both living documents that undergo constant modification during the project.

12.2.2 Functional Specification

A key characteristic of a reusable design is a pedigree of documentation that enables subsequent users to effectively use it. The requirements for a functional specification are outlined in Chapter 4. This specification forms the basis for this pedigree of documentation, and includes:

- Block diagrams
- Functional specification
- Description of parameters and their use

- Interface signal descriptions
- Timing diagrams and requirements
- Verification strategy
- Synthesis constraints

In addition to the above basic functional information, it is quite useful to keep the functional specification as a living document, which is updated by each user throughout its life. For each use of the block, the following information would be invaluable to subsequent generations of users:

- Project it was used on
- Personnel on the project
- Verification reports (what was tested)
- Technology used
- Actual timing and area results
- Revision history for any modifications

12.2.3 Project Plan

The project plan describes the project from a management perspective and documents the goals, schedule, cost, and core team for the project. Table 12-1 describes the contents of a typical project plan.

Table 12-1   Contents of a project plan

Part	Function
Goals	Describes the business reasons for developing the macro and its key features and benefits, including the technical goals that will determine the success or failure of the project.
Schedule	Describes the development timeline, including external dependencies and risks. The schedule should contain sufficient contingency time to recover from unforeseen delays, and this contingency should be listed explicitly.
Cost	Describes the financial resources required to complete the project: headcount, tools, NREs, prototype build costs.
Core Team	Describes the human resources required to complete the project: who will be on the team, who will be the team leader.

CHAPTER 13   *Implementing a Reuse Process*

This chapter addresses requirements for establishing reuse processes within a company. These requirements include tools, process inventories, macro libraries, and pilot projects. Topics in this chapter include:

- Key steps in implementing a reuse process
- Dealing with legacy designs

13.1 Key Steps in Implementing a Reuse Process

The following activities are key steps in implementing a design reuse process:

1. Develop a reuse plan.

   The first step in developing a reuse process is to develop a plan for establishing a reuse process. In particular, it is useful to determine the resources required to establish such a process.

2. Implement reuse training.

   Successful implementation of design reuse requires that design for reuse be an integral part of technical and management training within the company.

3. Inventory tools and processes.

   The next step in developing a design reuse process is to assess the design tools and methodologies already in place. A robust high-level design methodology that uses up-to-date synthesis and simulation tools is a prerequisite for developing a reuse methodology. Good project management practices and documented processes as outlined in the previous chapter are also required.

4. Build up libraries.

   Design reuse can begin with small designs. Build or purchase libraries of relatively simple components and use them in current design projects. Track the effectiveness of using these components and examine any problems that arise from their use.

5. Develop pilot projects.

   Develop pilot projects both for developing reusable designs and for reusing existing designs. This could involve existing internally-developed designs or a macro purchased from a third party.

   These pilot projects are the best way to start assessing the challenges and opportunities in design reuse. It is essential to track and measure the success of design reuse in terms of the additional cost of developing reusable designs and the savings involved in reusing existing designs.

6. Develop a reuse attitude.

   Design reuse is a new paradigm in design methodology. It requires additional investment in block development and produces significant savings in subsequent reuse of the block in multiple designs. This paradigm shift requires a change in attitude on the part of designers and management. Shortcuts in the name of "time to market" are no longer acceptable; the long-term cost to the organization is simply too high.

   Design reuse is the single most important issue in system-on-a-chip designs. Unless they consist almost entirely of memory, million-gate designs cannot be designed from scratch and hope to make their time-to-market and quality requirements. Reuse of previously designed and verified blocks is the only way to build robust million-gate chips in a reasonable amount of time.

   The most important step in developing a design reuse process is to convince the management and engineering team to adopt *the reuse attitude*: that the investment in design reuse is the key to taking advantage of the remarkable silicon technology currently available.

13.2 Dealing with Legacy Designs

Legacy designs—those designs we wish to reuse but were not designed for reuse—present major challenges to the design team. Often these designs are gate-level netlists with little or no documentation. The detailed approach to a specific design depends on the state of design. However, there are a few general guidelines that are useful.

13.2.1   Recapturing Intent

The most difficult part of dealing with a legacy design is recapturing the design intent. With a good functional specification and a good test suite, it is possible to fix, modify, or redesign a block relatively quickly. The specification and test suite fully define the intent of the design and give objective criteria for when the design is functioning correctly.

If the specification and test suite are not available, then the first step in reusing the design must be to recreate them. Otherwise, it is not possible to modify the design in any way, and some modification is nearly always required to port the design to a new process or to a new application.

The problem with recreating the specification and test suite, of course, is that these activities represent well over half of the initial design effort. Almost none of the benefits of reuse are realized.

Thus, if the specification and test suite exist and are of high quality, then reuse is easy, in the sense that even if a complete redesign is required, it will take a fraction of the cost and time of the original development. If the specification and test suite do not exist, then reuse of the design is essentially equivalent to a complete redesign.

13.2.2   Using the Design As-Is

In spite of the observations in the above section, some unfortunate design teams are required to try to reuse existing designs, usually in the form of netlists, for which documentation and testbenches are mostly nonexistent. In such cases, most teams attempt to use the design as-is. That is, they attempt to port the design to a new process without changing the functionality of the circuit in any way.

Formal verification is particularly useful in this scenario because it can prove whether or not modifications to the circuit affect behavior. Thus, synthesis can be used to remap and reoptimize the design for a new technology, and formal verification can be used to verify the correctness of the results.

13.2.3   Retiming

For some designs, the above approach does not provide good enough results. In these cases, behavioral retiming may be an effective solution. Behavioral retiming can automatically change the pipelining structure of the design to solve timing problems. Again, formal methods are used to prove that the resulting functionality is correct.

13.2.4 Tools for Using Legacy Designs

A large investment was probably made in legacy designs that are still very valuable macros. For example, a design team might want to reuse a macro developed for QuickSim II. ModelSim-Pro allows the team to simulate a VHDL and/or Verilog design that contains instantiations of QuickSim II macros.

If the design team wants to use a VHDL legacy design within a Verilog design (or vice versa), the ModelSim single-kernel architecture allows reuse of that macro within the context of the entire design.

13.2.5 Summary

Reusing legacy designs should definitely be the last aspect of design reuse implemented as part of establishing a design reuse methodology. Developing the processes for designing for reuse and for reusing well-designed blocks provides dramatically more benefit than attempting to reuse designs that were not designed with reuse in mind.

Glossary

ATPG – Automatic Test Pattern Generation

BFM – Bus Functional Model

BIST – Built-In Self Test

CBA – Cell Based Array

FSM – Finite State Machine

HDL – Hardware Description Language

ISA (ISS) – Instruction Set Architecture (Instruction Set Simulator); used interchangeably for an instruction set executable model of a processor

LMG – Logic Modeling Group

RTL – Register Transfer Level

SoC – System-on-a-Chip

SWIFT – Software Interface Technology

VFM – Verilog Foundary Model

VMC – Verilog Model Compiler

VSIA – Virtual Socket Interface Alliance; homepage www.vsi.org

Bibliography

Books on software reuse:

1. *Measuring Software Reuse*, Jeffrey S. Poulin, Addison Wesley, 1997.

2. *Practical Software Reuse*, Donald J. Reifer, Wiley, 1997.

Formal specification and verification:

1. http://www.ececs.uc.edu/~pbaraona/vspec/, the VSPEC homepage.

2. *Formal Specification and Verification of Digital Systems*, George Milne, McGraw-Hill, 1994.

3. *Formal Hardware Verification*, Thomas Kropf (ed.), Springer, 1997.

Management processes:

1. http://www.sun.com/sparc/articles/EETIMES.html, a description of the UltraSPARC project, mentioning construct by correction.

2. *Winning the New Product Development Battle*, Floyd, Levy, Wolfman, IEEE.

Books and articles on manufacturing test:

1. "Testability on Tap," Colin Maunder et al, IEEE Spectrum, February 1992, pp. 34 37.

2. "Aiding Testability also aids Testing," Richard Quinell, EDN, August 12, 1990, pp. 67-74.

3. "ASIC Testing Upgraded," Marc Levitt, IEEE Spectrum, May 1992, pp. 26-29.

4. Synopsys *Test Compiler User's Guide*, v3.3a, 1995.

5. Synopsys *Test Compiler Reference Manual*, v3.2, 1994.

6. Synopsys *Certified Test Vector Formats Reference Manual.*

7. *Digital Systems Testing and Testable Design*, M. Abromovici et al, Computer Science Press, 1990.

8. *The Boundary-Scan Handbook*, Kenneth Parker, Kluwer Academic Publishers, 1992.

9. *The Theory and Practice of Boundary Scan*, R. G. "Ben" Bennetts, IEEE Computer Society Press.

10. *Testability Concepts for Digital ICs*, Franz Beenker et al, Philips Corp, 1994.

11. "A Comparison of Defect Models for Fault Location with IDDQ Measurements," Robert Aitken, *IEEE Design & Test*, June 1995, pp. 778-787.

Books and articles on synthesis:

1. "Flattening and Structuring: A Look at Optimization Strategies," Synopsys Application Note Version 3.4a, April 1996, pp. 2-1 to 2-16.

2. *VHDL Compiler Reference Manual*, Synopsys Documentation Version 3.4a, April 1996, Appendix C.

3. *DesignTime Reference Manual*, Synopsys Documentation Version 3.4a, April 1996.

4. "Commands, Attributes, and Variables," Synopsys Documentation Version 3.4a, April 1996.

Index